Field to Lab

NIPA® GENX ELECTRONIC RESOURCES & SOLUTIONS P. LTD.
New Delhi-110 034

Field to Lab
A Comprehensive Handbook for Agricultural Extension

Sapna Jarial
Associate Professor
Department of Agricultural Economics and Extension
School of Agriculture
Lovely Professional University
Phagwara, Punjab

NIPA® GENX ELECTRONIC RESOURCES & SOLUTIONS P. LTD.
New Delhi-110 034

**NIPA® GENX ELECTRONIC
RESOURCES & SOLUTIONS P. LTD.**

101,103, Vikas Surya Plaza, CU Block
L.S.C. Market, Pitam Pura, New Delhi-110 034
Ph : +91 11 27341616, 27341717, 27341718
E-mail: newindiapublishingagency@gmail.com
Website: www.nipabooks.com

For customer assistance, please contact
Phone: + 91-11-27 34 17 17
Fax: + 91-11-27 34 16 16
E-Mail: feedbacks@nipabooks.com

ISBN: 978-81-19103-21-8

Composed and Designed by NIPA®.

Book Cover Design by Mr George Emanuel. Email: georgeemanual962@gmail.com

Preface

This laboratory handbook was created for agro-industrial attachment (AIA), and regular B.Sc. and M.Sc. agriculture and allied sectors extension courses. The RAWE primarily assists students in understanding the rural situation, the status of technologies adopted by farmers, farmer problems, and the skills and attitude of working with farm families for overall rural development. This book will assist students with step-by-step procedures for performing experiments on group exercises, participatory rural appraisal tools and techniques, presentation exercises, handling of audio-visual aids, writing scripts for radio and television, etc. This guidebook will benefit students, instructors, and policymakers, with the hope that we will always be open in our pursuit of positive change. For a wider application of field techniques, various exercises are described, and different types of equipment's standard operating procedures are explained. Methods are drawn from different resources. The author solicits suggestions for improving the manual.

Author

Contents

Preface .... *v*

1. To Perform Group Discussion Exercises 1
   - Learning outcome 1
   - What is the group? 1
   - Group Discussion 2
   - Why Group Discussion? 2
   - Group Discussion scoring tips 2
   - Case based group discussions 4
   - Procedure 4
   - Types of group discussion 5
   - How to conduct a panel discussion 6
   - Workshop 7
   - How to conduct a workshop? 7
   - Brainstorming 7
   - Seminar 9
   - Symposium 9
   - Steps to conduct symposium 9
   - Forum 10
2. To Study the Handling and Preparation of Audio-Visual Aids 13
   - Learning outcome 13
   - Classification of AV Aids 14
   - Activity 27
3. To Study Preparation of Extension Literature 29
   - Learning outcome 29
   - Preparation of extension literature 29
   - Steps to create material 30
   - Activity 33

4. To Study About Presentation Skills Exercise 35
Learning outcomes 35
What is presentation? 35
Oral message development 35
Features of Good Speech 36
Preparation work for speech 36
Methods for using visual aids 38
Presentation of research Paper 38
Parts of Presentation 38
Content of the Poster 40
Activity 42

5. To Study Micro Teaching Exercise 43
Learning outcome 43
Definition 43
Steps for micro teaching 44
Activity 46

6. To Visit the Village for Understanding the Problems Encountered by Villagers or Farmers 49
Learning Objectives 49
Techniques to improve rate of response 51
Questionnaire Construction 54
Activity 58

7. To Study Organization and Functioning of District Rural Development Agency and Various Development Department at District Level 59
Learning Objectives 59
Activity 62

8. To Visit NGOs to Learn from Their Experience in Rural Development 65
Learning outcomes 65
How to register an NGO 65
Activity 69

9. To Study Participatory Rural Appraisal Techniques with Their Application in Village Development 71
   - Why is PRA used? 71
   - How is PRA Conducted? 71
   - Materials and Equipment 71
   - Principles of PRA 71
   - Activity 81

10. To Visit the Community Radio Television Studio for Understanding the Process of Programme Production 83
   - Learning outcomes 83
   - Introduction 83
   - Types of radio 83
   - CRS - Criteria for Site Selection 84
   - Bonus Material 87
   - Activity 95

11. To Study Script Writing in Print and Electronic Media 97
   - Learning outcome 97
   - What is Print Media? 97
   - What is Electronic Media? 97
   - Considerations for writing for print media 97
   - Activity 109

12. To Study the Script Development for Radio and Television 111
   - Learning outcomes 111
   - Different types of formats 112
   - Activity 122

13. References 125

1
To Perform Group Discussion Exercises

Learning outcome

The purpose is to enable the student to develop deeper knowledge understanding capabilities and attitudes in the context of group discussion.

What is the group?

A social group is a unit of two or more person in the reciprocal interaction of communication with each other.

Stages of Group Formation

Bruce Tuckman presented a model of five stages.

Orientation (Forming Stage)

During this stage, group members come together and begin to establish relationships with each other. They also clarify their goals and objectives for the group.

Power Struggle (Storming Stage)

As the group begins to work together, conflicts and differences in opinions may arise. This stage is characterized by power struggles, disagreements, and a lack of cohesion among group members.

Cooperation and Integration (Norming Stage)

In this stage, the group begins to resolve conflicts and work towards a common goal. Members begin to cooperate with each other, establish norms and rules, and develop a sense of unity.

Synergy (Performing Stage)

Once the group has established a clear sense of direction and unity, they can focus on achieving their goals. This stage is characterized by high levels of productivity and synergy among group members.

Closure (Adjourning Stage)

This stage occurs when the group completes its task or project. Members may feel a sense of loss as the group disbands, but they can reflect on their accomplishments and lessons learned.

Group Discussion

Group discussion (GD) is an activity where individuals congregate to analyse and solve common problems.

Why Group Discussion?

To ascertain potential personality traits of adaptability, flexibility, leadership, teamwork ability. It helps in choosing charismatic extroverts.

Group Discussion scoring tips

Initiate - Introduce the topic first.

Lead - Channel discussion and urge others.

Organize - Structure discussion and keep GD on topic. Redirect the group if it strays.

Conclude - List key points to create a broad conclusion.

Example:

How to initiate

Dear friends, we have a great topic to talk about. I'd like to talk a lot about this subject. I'm glad to observe that everyone seems just as willing to talk about their ideas. So, I'd like you to tell me one at a time what you think about this topic, so that we can find things we agree on.

- Who will start?
- Who would like to introduce the topic to the group?

How to lead?

Due of limited time, please be concise and explain only the most important things in this GD.

In the first round of discussions, avoid specificity. Detailed discussion is possible if time permits.

How to argue?

I agree with some of your comments but disagree with one. And I would like you to explain me further on that point. I think that....

How to conclude?

Mr. XX introduced the themes and outlined 4 critical issues. Mr. YY clarified the relations between the important issues variables. We all agreed after long conversations on the points:

A...
B....
C...
D...

Understand cross cultural communication. Use clear voice tone speech language slow pace.

Non-verbal communication is more important than verbal communication informs of body language sitting posture proximity, smile and facial expressions.

GD can be topic and case based. Topic based GD can be divided into 3 themes: Factual, controversial, and abstract.

Topic based group discussion- Some Evergreen topic for example: global warming or greenhouse effect; brain drain; Internet boon or bane; women empowerment; cybercrime etc.

1. **Factual themes-** Facts are everyday stuff an average person knows. On the other hand, socioeconomic themes are specific factual topics. These can be recent or timeless. A fundamental discussion topic allows a candidate to show s/he's awareness and sensitivity to their environment. For example, the education policy of India, tourism in India, state of the aged in the nation.
2. **Controversial themes-** These are contentious subjects; aimed to spark debate. In GD, these topics are discussed loudly. The purpose of giving such a topic is to see how mature the candidate is by keeping their temper in check and debating reasonably and logically without getting personal and emotional—for example, caste-based reservations, gender inequality, women drivers etc.
3. **Abstract themes-** Abstract topics are about intangible things. These topics are not given often for discussion, but their possibility cannot be ruled out. These topics test your lateral thinking and creativity. For example, a is an alphabet, the number 10 etc. Intangible things constitute abstract themes. These aren't typically discussed, yet they're possible. These themes require creative and lateral thinking, e.g., number 13, alphabet z, etc.

Case based group discussions

The case study simulates a real-life situation. You'll be given situation details and asked to handle it as a group. In the case study, there are no incorrect answers or perfect solutions. Case studies aim to have you consider the subject from many perspectives.

Harvard Business School, U.S.A. developed this technique. It is used as a supplement to the lecture method.

Procedure

- The trainee receives textual information and complex organisational scenarios. A case study can be 10 to 200 pages depending on the organisation's difficulty.
- Case studies often end with questions.
- Longer case studies provide adequate investigation information, while shorter ones require study.
- The learner then evaluates the scenario and suggests solutions.

Discussion killers in GD

These words stop creative discussion.

That's crazy.
Impractical.
Let's get back to
You are wrong…
You do not know what you are talking about...
No that's not what I'm talking about

Igniters of GD

These words are used to encourage others in a positive way.
That's good!
I agree…
That's a great idea.
I'm glad you brought that up…
You are on the right track…
I never thought of that…
We can do a lot with that idea…
Really good, anyone else?

Tips for ensuring success in Group discussion

Don't		
Destructive body language biting nails, shaking legs, picking nose, playing with pen, do not lean back in chair with hands clasped behind head it indicates arrogance or disinterest.	Indifference, spectator behaviour. Cannot sum up or finish. Neither agrees with nor is open to different ideas. Afraid and doesn't say much	Self-centred fails to involve everyone. Initiate and generate the discussion. Do not be too eager to talk a lot
Do		
Be assertive (not aggressive) Avoid blocking. Use proper language. Avoid attention seeking. Participate fully and actively Don't act as a facilitator. Avoid individual conversation.	Be a patient but active listener. Seek clarification. Respecting others' opinions doesn't imply inferiority or superiority. Be precise. Summarise.	Be analytical and fact oriented. When necessary, intervene strongly but gently. Show leadership ability or harmonise. Maintain eye contact. Smile, it exhibit light-hearted frame of mind.

Types of group discussion

1. Small group discussion
2. Panel discussion

3. Workshop
4. Brainstorming
5. Symposium
6. Forum
7. Seminar

1. **Small group discussion-** Small group techniques are the ways that a group works together to get things done. Small GD lets presenters tell participants about the topic or idea they want to discuss as a group. A small group discussion follows democratic rules and lets everyone share many ideas that others can talk about and think about.

2. **Panel discussion-** Are talk about things people care about, but only a few people talk about them in front of an audience. It is meant to give a group a chance to hear from several people who know a lot about a particular issue or topic. At the end of the conversation, people can also participate.

Members in the panel discussion constitute of moderator, panellists, audience, and instructor.

Panellist-there are 4 to 10 panellists in the discussion. Members sit in a semi-circle shape in front of the audience.

The moderator sits in the middle of the panelist. All panellists should have mastery over the subject matter.

Audiences- are allowed to put questions and seek clarifications, and they can put forward their perspective.

How to conduct a panel discussion

1. Identify the goal and purpose of your panel discussion
2. Invite experts to become panellist
3. Select an invite a moderate
4. Determine the rules for the panel discussion

Cautions

The moderator and panellists determine this technique's success.

One or two members should not monopolise the discussion.

This procedure requires careful planning. Allow enough time. Panels shorter than 30 minutes may not succeed. While group interest is high, end the panel discussion.

Workshop

A workshop is a period of discussion or practical work on a particular subject in which a group of people share their knowledge or experience. It could be theoretical, or discussion-based; not limited to practical skills and it is subject-specific. The duration of the workshop varies from a long meeting of one day to several weeks involving all the delegates. Here, the problems being discussed are considered by delegates in small private groups. There must be a planning session where all are involved in the beginning. During the workshop, considerable time for work sessions should be given. There must be summarising and evaluation sessions at the close. In the end, a report for a program for future action application, a visual or any other material object is expected.

How to conduct a workshop?

1. Select the theme, formulate a team, and invite the experts.
2. At planning, the venue, workshop topic, duration, boarding, and lodging are to be discussed.
3. Concept notes should be circulated with participants, with clear expectations.
4. Enough time for travel, visa arrangements, and hotel accommodations should be made.
5. Conference room/hotel room should be booked for the workshop.
6. At the workshop, proper groups should be made.

Brainstorming

Brainstorming is storming your mind with lots of ideas. Brainstorming is the quick, non-critical creation of ideas in a particular period. Information is sorted, synthesised, and discussed. It identifies issues.

Brainstorming is a group creativity process that gathers participants' spontaneous thoughts to solve an issue. Alex Faickney Osborn popularised the word in 1953. in his book-

"Applied Imagination". Brainstorming is a training discussion strategy. It's a potentially dynamic teaching method, but its success depends on how well it's organised and run.

Types of Brainstorming - Individual brainstorming and Group brainstorming

Individual Brainstorming

When you brainstorm alone, you may be free and creative without worrying about others' egos or opinions.

Steps for individual brainstorming

- Find a quiet spot to reflect.
- Focus on the problem, minimise distractions,
- Use mind maps to list and arrange ideas,
- Then choose the best idea
- Develop idea.

Group Brainstorming

The whole experience and creativity of everyone on the team are a plus.

When one member gets stuck on an idea, the creativity and experience of another member can help generate the idea and to move to the next stage.

Steps for group brainstorming

1. Prepare the group
    - Set up a comfortable place for the meeting.
    - Include people from diverse fields and thinking styles.
    - One person should record session thoughts.

2. Present the problem
    - Identify the issue
    - Set requirements.
    - Give them time at the start of the session to write down their ideas.
    - Then share the ideas.
3. Guide the discussion
    - Start a group discussion to build on others' ideas and create new ones.
    - Encourage creativity and as many ideas as possible, practical or not.
4. Analyzing the ideas
    - Brainstorming exercise will generate many ideas.
    - Analyze and choose the finest ideas.

Rules for Brainstorming

- No criticism is allowed during brain storming.
- Quality of ideas is important.
- Wildness is good. Crazy ideas are welcome.
- Many a times craziest ideas turn out to be the best ones.

Seminar

Seminars are a form of academic training at academic institutions or commercial or professional organisations. It brings small groups together for recurrent meetings on a certain subject in which everyone present is asked to participate. It is conducted through Socratic dialogue with the seminar leader or lecturer or formal research presentations.

So, the seminar is a structured method of group conversation, which lets professionals examine certain issues. The discussion leader introduces a topic. In a seminar, participants present research-based discussion papers. Individual thoughts can be given directly or indirectly related to the topic and everyone is requested to participate. Audience members discuss an unanswered question.

Two or more plenary sessions may be held in seminar. This strategy combines many opinions. Finally, conclusions and future action recommendations are made.

Symposium

A symposium is a group of talks, speeches, or lectures from different people on different parts of the same issue.

Oxford Advanced Learner's Dictionary describes it simply as "a small conference".

Symposiums usually focus on one issue rather than a more prominent theme. A group of experts will get together to share their ideas and papers. It is typically completed in a single day.

A symposium is a group of talks, speeches or lectures presented by several individuals on various phases of a single subject problem.

Steps to conduct symposium

The identified problem must have practical components amenable to the sessions.

A moderator controls time and subject matter. The time for the talk should be within 20 minutes, and the total time should be within one hour. In a symposium, the audience may or may not participate.

Duties of Moderator / Chairman

- S/He should meet the symposium speakers early to agree on a logical order.
- Outline the topics and agree on order and time.
- Before the meeting, s/he should examine the above items.
- At the meeting, s/he should describe the sequence of speakers and create enabling environment for listening and thinking.
- Inform the group of the symposium's protocol, including speakers' and the group's roles during and after.
- Introduce each speaker and their qualifications as they talk.
- Depending on the procedure, perform extra functions such as summarising or following up.
- Allow each speaker a quick clarification.
- Each speaker can ask other questions.
- Use direct audience questions.

Duties of speakers

Attend moderator-organized "planning meetings".

- They should create brief, well-organized presentations in the specified time.
- Present the subject clearly in allotted time.

Caution

- Care must be given in choosing a topic and breaking it into meaningful pieces.
- Name the moderator and speakers carefully.
- Unbiased, logical speakers should be chosen.
- The moderator isn't a speechmaker but an interpreter. S/He should handle it well.
- Plan to limit speech length and enforce it.
- Logical, precise presentations are vital to a productive symposium, but the larger group's aim must also be fulfilled.

Forum

It is also called as "Question – answer period". It is a discussion period that may follow any one of the above methods of presentation. It includes a question-and-answer opportunity for audience members. The forum allows the audience to clarify and ask questions.

Activity

1. Perform group discussion on innovative technologies in agriculture.

Notes

2

To Study the Handling and Preparation of Audio-Visual Aids

Learning outcome

In this exercise we will learn about handling of:

1. 18-135 mm lens
2. 70-200 mm zoom lens
3. 8.3.5 mm DSLR lens
4. Babylight
5. Computer
6. DSLR camera
7. Field recorder
8. Mic rod
9. Power adaptor
10. Projector
11. SD card 32 GB
12. Speaker
13. Tripod
14. Projector screen

Research on how people learn has shown that, of the five senses, people learn most through their sight. Some have said that visual learning makes up as much as 80% of a person's life. Learners remember 10% of what they read, 20% of what they hear, 30% of what they see, and 50% of what they hear and see. As said, visual media has a big impact.

Audio-Visual Aids are those devices using which communications of ideas between persons and groups in various teaching and training situations is helped. These are also termed as multi-sensory materials

- *Edgar Dale*

Audio-Visual aids are devices that appeal to the eyes and ears of the learners and the materials used are light or sound for presenting the information.

Audio-visual aid is anything the audience can see and hear that helps the speaker to get his message across to the audience.

Why use them?

Visual and audio-visual aids are important part of learning process and aids in helping people learn immensely.

Classification of AV Aids

We can group aids into three types according to the senses they appeal to:

Aid	**Senses**
Visual Aids	appeal to the sense of sight
Audio-Visuals Aids	appeal to the senses of sight and hearing
Audio Aids	appeal to the sense of hearing

Definition

	Babylight: They are a light source of varying size and intensity and can be used as a spotlight or a flood light.
	Computer: an electronic device for storing and processing data, typically in binary form, according to instructions given to it in a variable program.
	DSLR Camera digital single-lens reflex camera (digital SLR or DSLR) is a digital camera that combines the optics and the mechanisms of a single -lens reflex camera with single-lens reflex camera with a digital imaging sensor.

	18-135 mm lens: is a high-performance high-magnification zoom lens equipped with an Image Stabilizer and has been developed for digital SLRs compatible with EF-S lenses.
	3.5 mm DSLR Lens: is a sort of prime lens that is used to take pictures of near object with a DSLR camera.
	70-200 mm zoom lens: gives a narrow field of view along with a fast aperture, it can easily remove distracting elements from the frame and then blur everything else remaining so that the subject stands out against the background, it can quickly change the field of view without having to physically move.
	Field Recorder is an equipment used for an audio recording produced outside a recording studio.
	Mic rods: are telescoping, normally with four to five sections, but sometimes up to seven. When extended, they can range in length from 6 to 12 feet, and even longer. The mic rods are an equipment widely used for reaching out to distant voices.

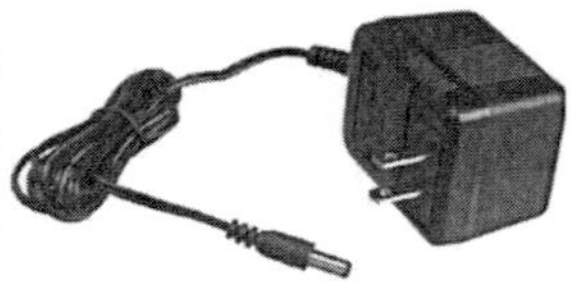	Power adaptor: is used to charge the camera`s battery.
	Projector: A projector or image projector is an optical device that projects an image (or moving images) onto a surface, commonly a projection screen.
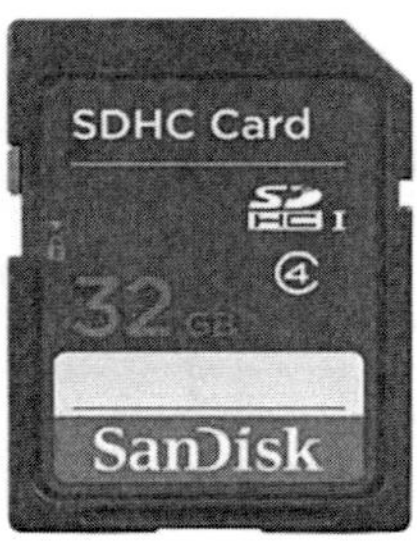	SD card 32 GB: is a solid-state drive is a solid-state storage device that uses integrated circuit assemblies to store data persistently, typically using flash memory, and functioning as secondary storage in the hierarchy of computer storage.
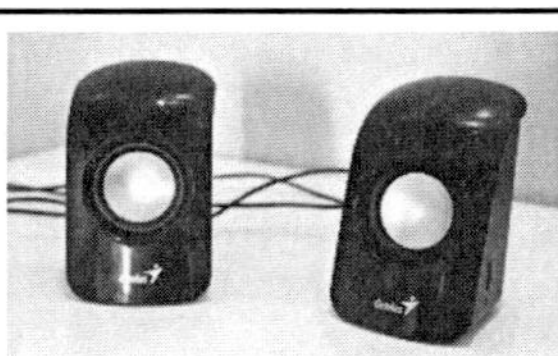	Speaker: is an equipment widely used in Agricultural Extension Education laboratory and various educational Institutes, research, and development sector.
	Tripod: is a portable three-legged frame or stand, used as a platform for supporting the weight and maintaining the stability of some other object. A tripod provides stability against downward forces and horizontal forces and movements about horizontal axes.
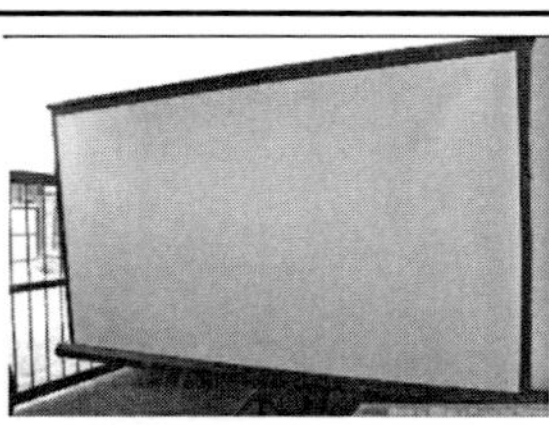	Projector Screen

Steps to turn on the computer

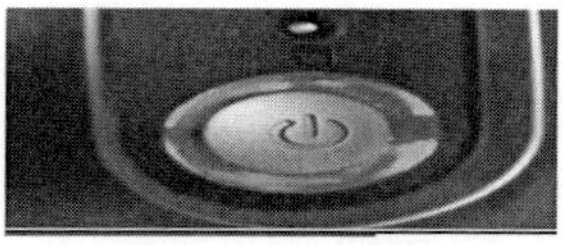	Press the 'ON' the inner front of the computer.
	Wait until the display produced on the screen. Input username and password (if necessary).
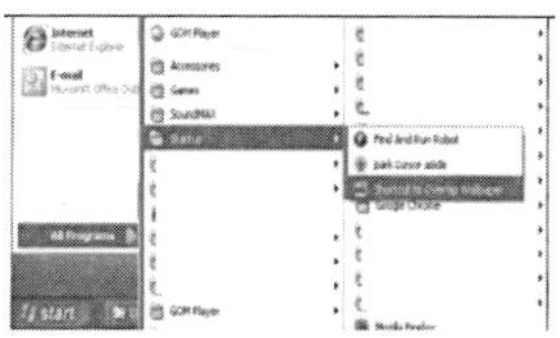	Click on 'START' and select 'PROGRAM'
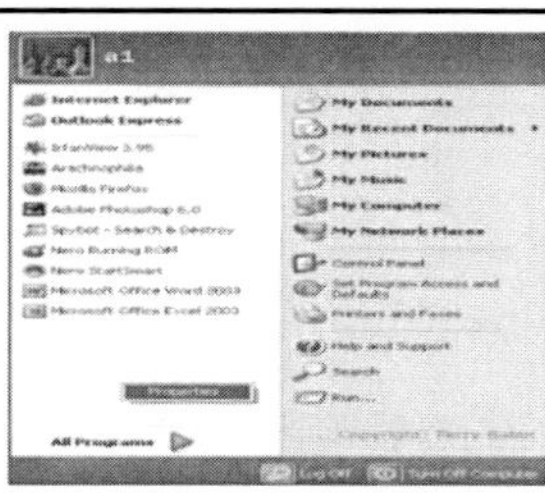	Select the program you want to use and start using the computer
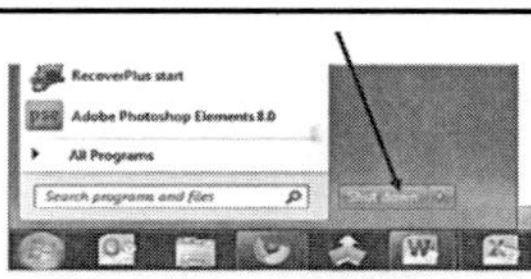	Click on 'START' and select 'Turn off COMPUTER' to end the use of computers.
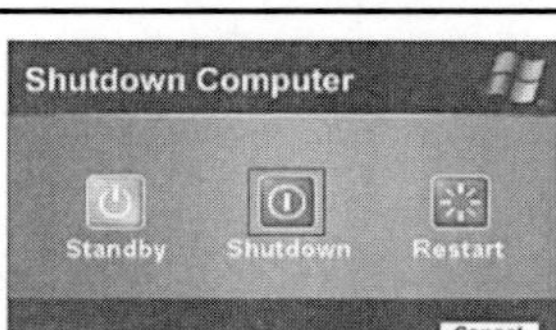	Select "Stand By" or "Shutdown" or "RESTART" computer

(Source Techiweb.com)

Steps to connect Speakers to Computer

If you own a laptop, you probably dislike the built-in speakers. Buying computer speakers will improve your laptop movie-watching and music-listening. Wireless or wired computer speakers are straightforward to set up on a PC or laptop.

1. Connecting Wired Computer Speakers

	Choose computer speakers. Most wired computer speakers should work with any laptop with a USB or speaker/headphone jack. Most computer speakers feature 3.5mm (1/8") audio input connectors, which fit into headphone jacks. These speakers need power.
	Arrange the speakers in your workspace. The back or bottom of most speakers is designated L(left) or R(right). If you have a power cable, be sure it can reach an outlet from wherever you install your speakers.
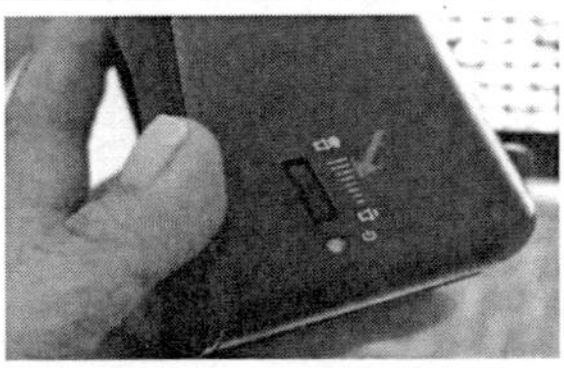	Turn the speaker volume to the minimum. To do this, turn the speaker's (one) volume dial to the left.
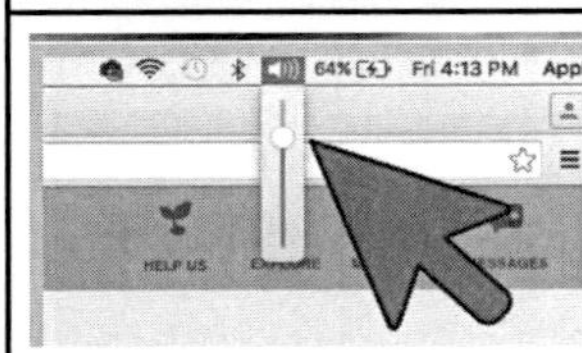	Adjust laptop volume to 75%. Click the taskbar sound icon (bottom right corner on Windows).
	Into the laptop plug the connector. Plug the USB or 3.5mm (1/8") audio cable into the laptop while it's on.If you're using a 3.5mm (1/8") connector, search for little port with headphones drawing or a speaker on your laptop. Don't use the microphone jack.
	Plugging in USB speakers may prompt system driver installation. Insert the disc that came with your speakers and follow the instructions. Turn on the Speakers. The speaker's On button is on the back of one of the speaker. Before turning on your speakers, plug in their power cable.

(Levine, 2022)

2. Connecting Wireless Computer Speakers (Bluetooth)

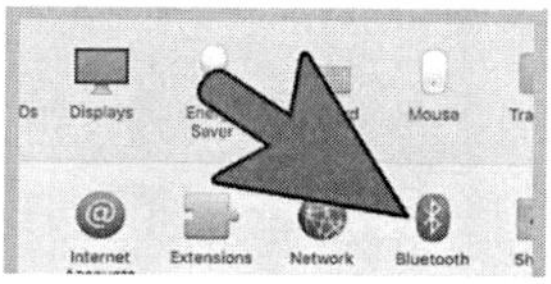	Check for Bluetooth in your laptop In Windows, press Win+X and select "Device Manager." Click "Laptop." Click "Bluetooth Radios" under Laptop to display Bluetooth devices. If you have Bluetooth, check this list.
	Locate your Bluetooth speaker(s). The plug speakers to power source. Speakers can be up to 30 feet away, but some devices may have lower ranges.
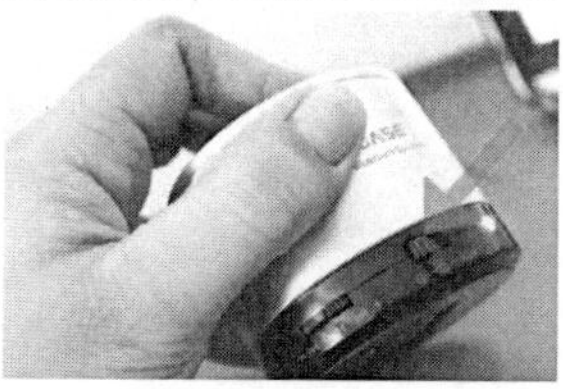	Turn on your Bluetooth speaker and make it discoverable. Many speakers have a button that must be pressed for a few seconds to enter "discovery" mode.
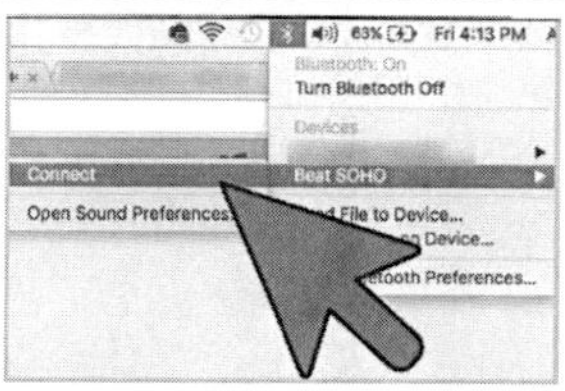	Pair your laptop with the Bluetooth speaker. This process depends on your operating system: Click the notification icon in Windows 8 or 10 to open the Action Center (near the clock). "Bluetooth," then "Connect," to find devices. Select the speakers to connect.
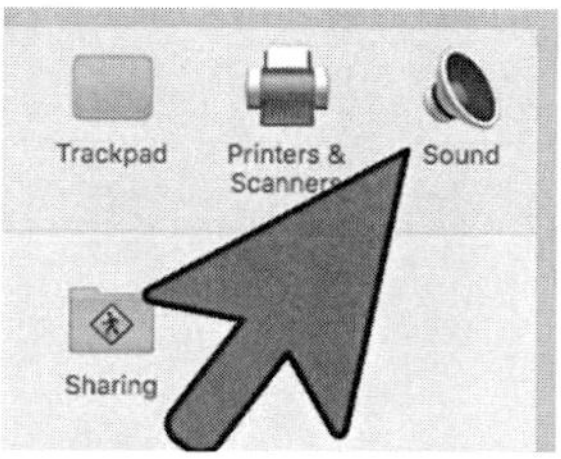	Configure your computer's speakers. Press Win+S and type control. Click "Control Panel," then select "Sound." On Playback tab > Bluetooth speaker and click "Set Default." Click OK.
	Test the audio

(Levine, 2022)

Steps to Mount Camera on Tripod

1. Check that your camera has a tripod mount. 2. Remove the camera plate from your tripod.	
3. Level the tripod. Adjust the legs so that they stand stably on the ground. Release the tensioners on the legs, and raise them to the height that you need them. You can technically attach a camera to the tripod before you set up the tripod – but your camera will be safer if you set up the base first.	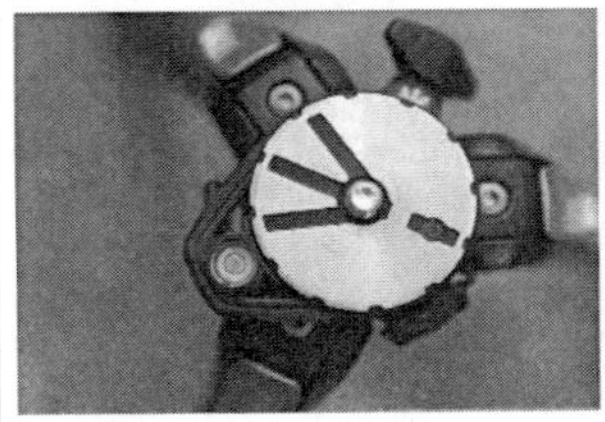
4. Mounting the Camera 5.Screw the camera onto the tripod. 6. Clamp the camera to the tripod. 7. Make sure that you're using the right camera plate.	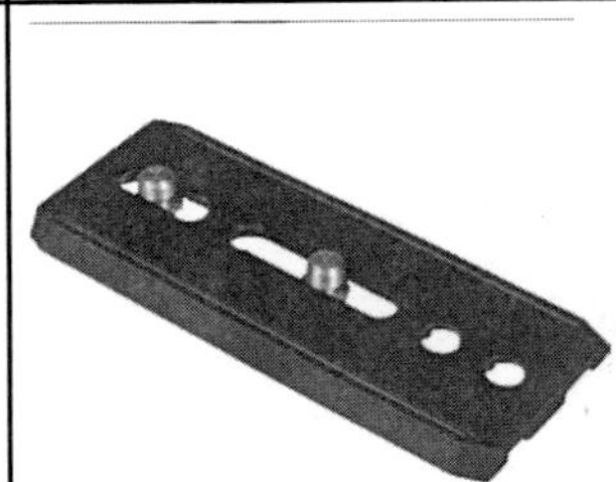
8.Hang your camera bag from the center column of the tripod.	
9. Don't attach the camera directly to the legs. Most professional-grade tripods are sold with the legs and tripod head packaged separately	

Steps to Use babylights

Film sets employed these inexpensive, easy-to-use lights. They can be used as a spotlight or flood light and vary in size and intensity. The lamp sits on a moving track behind a Fresnel lens and in front of a spherical reflector. The reflector sends light to the Fresnel lens, which forms a beam. All of this is housed in a heat-dissipating housing. A knob in the back of the housing moves the lamp/reflector team down a track to spot or flood the beam. Three fixed and one moveable brackets hold barn doors, speed rings, scrims, and other items. Finally, the lamp has an AC power connector and a standard-inch (16 mm) stand mount and yoke.	 **Fig. 1 :** Babylight (source: Red Head Kit – 3 Light 800W – Cameralight | Lighting solutions for Photographers & Videographers)

1. Plug the babylight cable into high power electricity point.
2. Take care, not to step onto the long wire of the babylights.
3. Allow them to cool off before packing.

Steps to take picture from DSLR Camera

	1. Charge batteries. You don't want to be out with your camera and find your battery dead.
	2. Insert a memory card. Buy extra memory cards for taking photos. Memory space needed depends on photo quality.

	3. Start your camera. This is the first camera skill you'll master. You'll forget where the on/off switch is soon. Next, manage aperture, shutter speed, and ISO similarly. More use makes your camera easier. You can alter settings without paying attention.
	4. Select a mode. Most cameras feature exposure modes. Aperture priority mode is popular. Popular modes include shutter priority and auto. This is especially true for inexperienced photographers who haven't discovered manual mode.
	AUTOMATIC Automatic mode lets the camera determine exposure. Unless you set it to no-flash auto mode, the flash will go off if needed. Simple point-and-shoot photography uses this. Program Mode Program mode sets shutter speed and aperture. This mode is useful for snapshots and when you're pressed for time. The camera chooses the exposure in this mode. This is regulated by the camera's programming and works for Canon and Nikon. Shutter Priority Mode The shutter speed determines the aperture. This option freezes/blurs motion. Shutter priority mode is useful for learning your settings. Shutter speed management is crucial. Don't use a slower shutter speed than necessary. This is vital for moving subjects. A quick shutter speed prevents camera shake blur. When a slow shutter speed is utilised and the camera moves, this happens. The photo will be blurry overall.
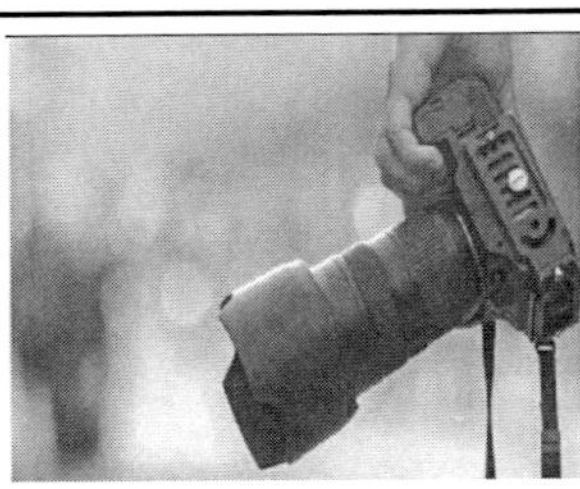	Aperture Priority Mode You choose the aperture while the camera chooses the shutter speed. If you want the background hazy or everything in focus (DoF), use this. This setting will help new photographers learn how to manage depth of field.

	Manual Mode You choose shutter speed and aperture in manual mode. Hold the +/- button and turn the dial to adjust. This exposure compensation control lacks precision. Learning to control the aperture and shutter speed is best. Auto ISO should never be used if you want to regulate exposure. Auto ISO implies the camera controls the exposure. The camera manages ISO regardless of aperture and shutter speed. The camera controls the exposure, too. It determines the setting and adjusts ISO.
	Scene Modes/Effects These modes are for special effects and scene optimization. If you want to learn camera, ignore these settings. They make photography seem easy. Easy photography isn't worth owning a camera for. Phones work. Scene modes might be confusing. Master your camera. Few parameters matter. Once you master them, you won't need scene or effect modes.
	Information Display Options Almost every digital camera has a "info" button. This button displays shooting information on the monitor. Learning to interpret the screen takes time. It's worth understanding. This will improve your photography. It's used to review photos.
	Image Quality Two image qualities are available. Your camera can save RAW or jpeg photographs. This is crucial. Learn how to save photos if you want to alter them or not.

(Domien, 2022)

Steps to Handling and Protection of Camera

Camera Body Cleaning	
	Remove the lens and other accessories from the camera. Turn off the camera before removing parts. Remove the camera's battery and strap. Press the lens release button to separate the lens.

<table>
<tr><td>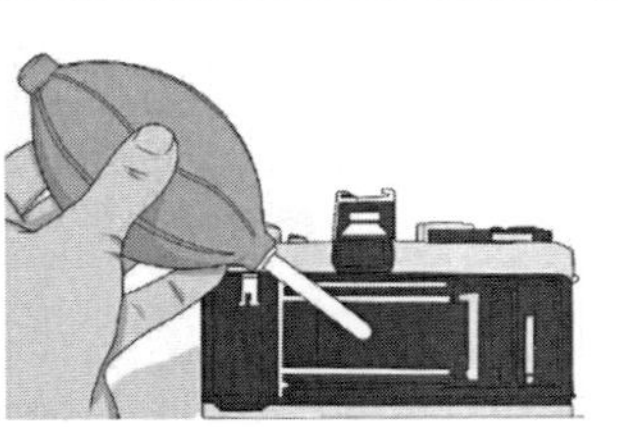</td><td>A blower is used to clean the inside of the film chamber. A squeezed blower puffs air. Open the film chamber and face it downward. Squeeze the blower to blow away any particles in the film chamber.</td></tr>
<tr><td>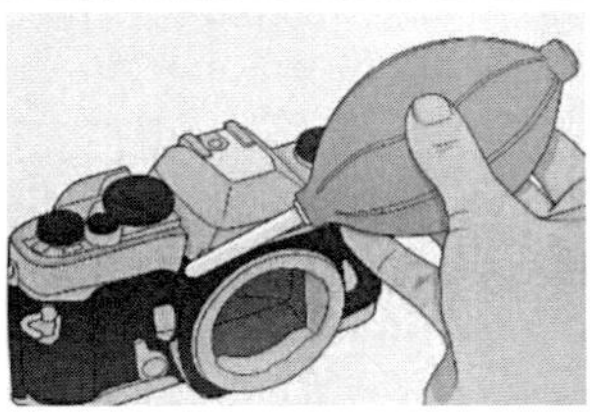</td><td>Use a blower to remove dust from the camera. Use the blower to remove dust from the camera's exterior. Do this before wiping the camera with a cloth, as any residual particles could leave scratches.</td></tr>
<tr><td>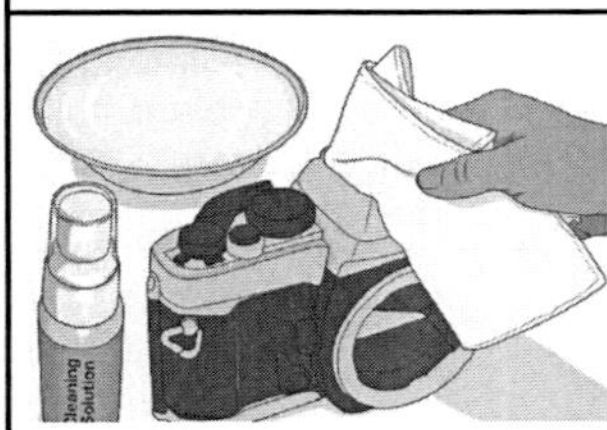</td><td>Wipe the camera with a wet cloth. Use camera-specific cleaning supplies. You don't want solution leaking onto the camera and soaking into cracks.

If you don't have a cleaning cloth, use a cotton swab or toothpick wrapped in cotton.</td></tr>
<tr><td></td><td>Remove difficult stains or tarnish with a pencil eraser. Rub the eraser on camera spots and grime. After using the eraser, use the blower or a lens brush to remove eraser crumbs.</td></tr>
<tr><td>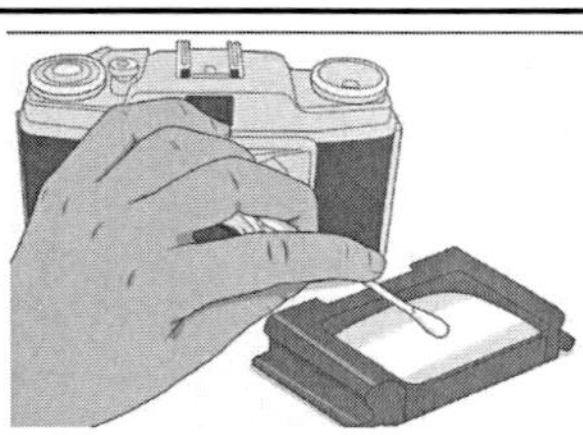</td><td>Wipe or wipe the camera's ground glass. You may reach ground glass by raising the little clasp behind the lens mount. Wipe down the ground glass with a cleaning solution-soaked cloth or cotton swab.

Use a blower to remove particles from ground glass before wiping it down.</td></tr>
<tr><td>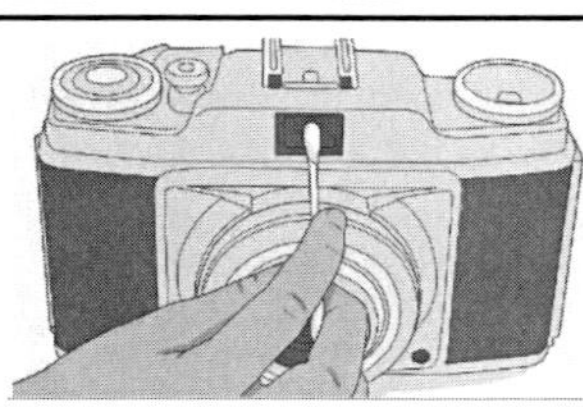</td><td>Use Cotton-swab to wipe the viewfinder in cleaning solution to remove smudges and oil.</td></tr>
</table>

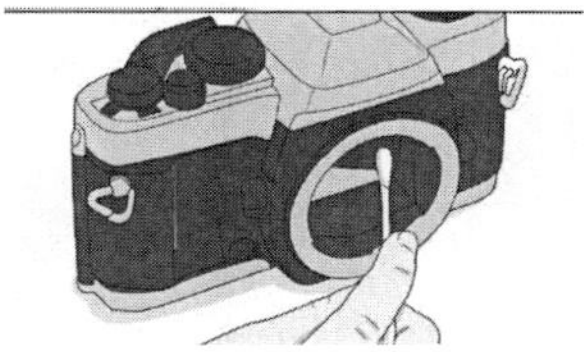	Clean the mirror with a cotton swab. The mirror is located behind the lens mount. Using cleaning solution, wipe a cotton swab across the surface of the mirror. Let it dry and check to make sure there aren't any streaks before you put the lens back on.
Lens-Cleaning	
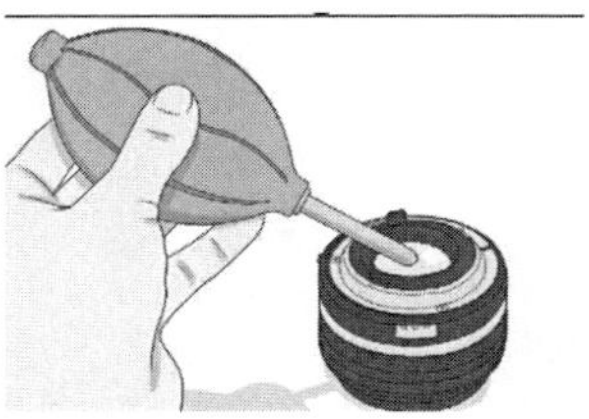	Clean your lens with a blower. Give your lens a couple squeezes to remove any dust. Then squeeze the blower close to your lens to blow away particles. Don't blow on your camera with your mouth. Saliva can harm a lens.
	Dust your lens with a lens brush. Use a camera-lens-specific brush to avoid damage. Fine lens brushes won't scratch glass. Move the lens brush in a circular motion to remove dust.
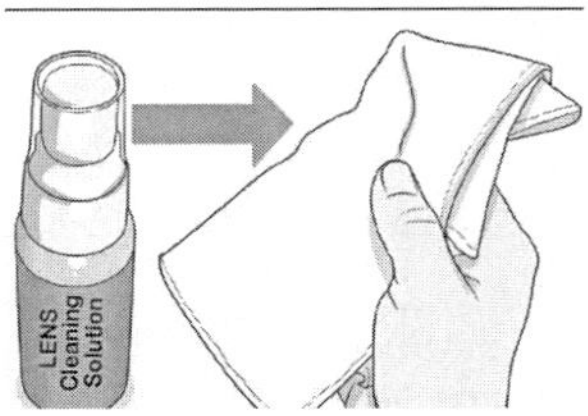	Clean lens with tissue or cloth to avoid scratches. Don't wipe your lens with tissue or a t-shirt. Never apply lens cleaning straight to your camera lens. Applying a lens cleaning directly to your camera lens can get liquid inside.
	Wipe lens with a cloth. Concentric circles prevent streaking. Don't push too hard; just enough to remove smudges and oil.
Camera and lens maintenance	
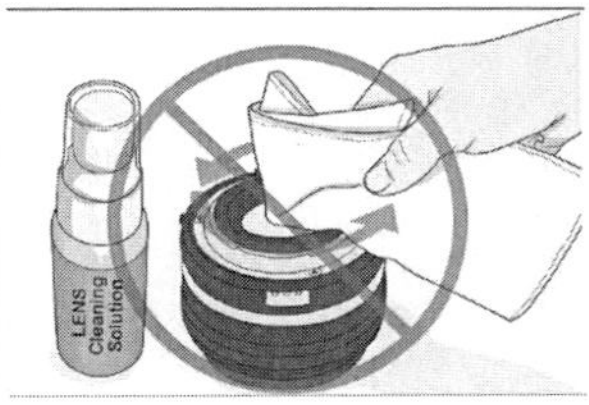	Don't overclean your lens. Overusing chemicals and physical contact can damage camera lenses. Avoid touching your camera lens when shooting images to reduce cleaning. Some dust is OK and won't affect the quality of your images.

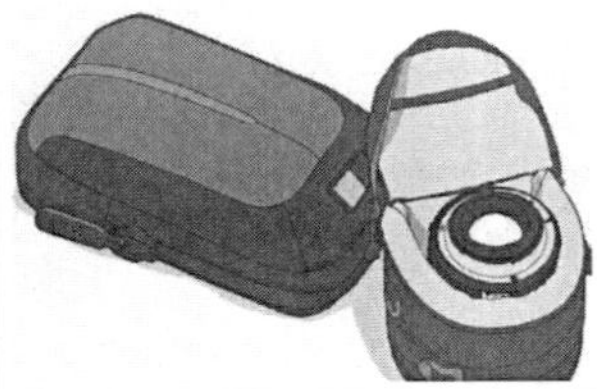	Always keep your camera and lens in a case.
	Protect your lens with a cap.

(Ryan, 2022)

Steps regarding Mics for field recording

There are various types of mics:

1. Lav mic
2. Shotgun mic
3. Dynamic mic
4. Digital recorder/field recorder

Type	**Details**
1. Lav mic 	The portable lav microphone is often used outdoors. Whether you're a reporter on the go or filming a YouTube travel vlog. They are small, easy-to-use. Being able to get clean audio from a pocket-sized device can be useful if you lack audio professionals. Plug it into your camera or voice recorder for sharp, clean audio. Lav mics work well if they're close to the speaker.
2. Shotgun mic 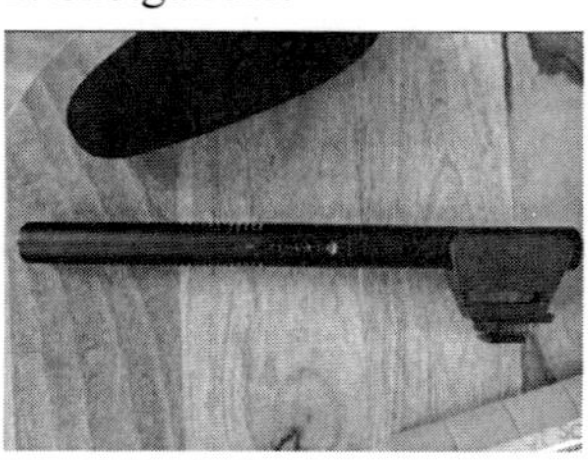	Shotgun mics are popular for outdoor recording. Shotgun mics look like a long stick with a microphone on the end. Its directional pickup technique implies it won't catch up much side sound. Therefore shotgun mics face the speaker directly. Rode VideoMic attaches to the camera and points at the speaker.

3. Dynamic microphone 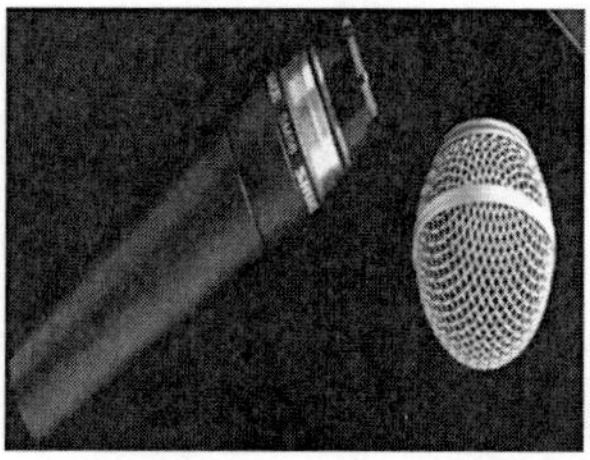	When a reporter passes the mic back and forth, they use a dynamic mic.
4. Digital field recorder 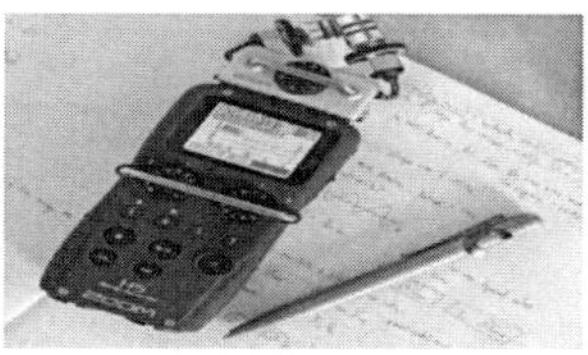	Digital recorders make good outside recordings. Zoom H1n, Zoom H5 are some examples. Pocketable and easy to use. On and record.
5. Studio condenser mic 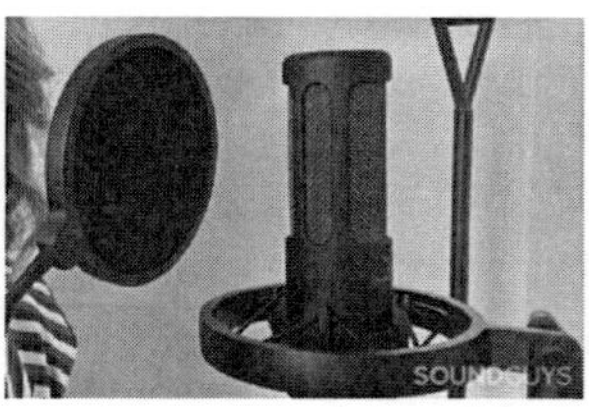	Your condenser microphone is praised for picking up everything. While this is desirable in a studio, most condenser mics can't isolate your subject in the field. Dynamic mics and lav mics let you better control what makes it onto field recordings. High-sensitivity studio condenser mics catch everything, from your foot changing to a distant dog barking. Condensers vary. Studio condensers aren't as rugged as field models. A good rule of thumb is if it appears like something a singer would use in the studio, it doesn't fit outside.

Activity

Explain the handling of field recorder.

Notes

3
To Study
Preparation of Extension Literature

Learning outcome

The purpose is to enable the student to develop deeper knowledge understanding capabilities and attitudes in the context of preparation of extension literature.

Extension materials are straightforward messages that encourage, educate, and improve farmers' understanding of an innovative, proven technology. Therefore, extension materials must encourage awareness and highlight the benefits and effective application of innovative technologies. Orally or through bulletins/brochures, fact sheets, posters, radio dramas, and videos, extension materials can suggest innovative technologies. Literature is the foundation of any instructional programme. Extension materials must persuade individuals to alter their behaviour.

Preparation of Extension Literature

1. Extension literature should be simple.
2. Before preparing extension materials, find out the priority needs of the target audience
3. Determine the most efficient format for addressing the target audience.
4. What changes do you anticipate?

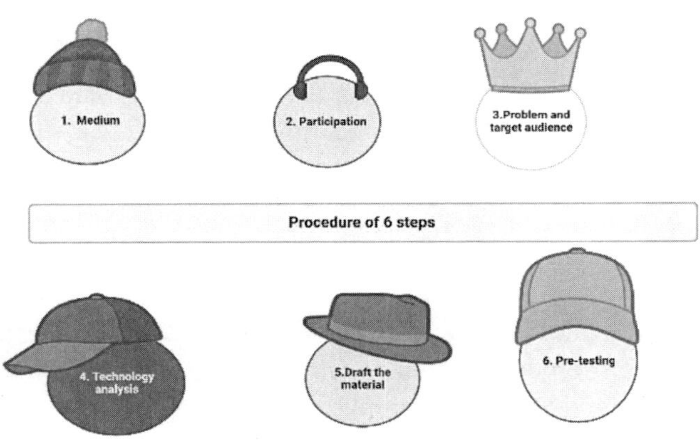

Picture 1: Procedure to create an extension material

Steps to create material

1. Choose the communication medium (written, radio, video etc).
2. Develop the content in participation with a variety of interested groups (farmers, other extension workers, scientists etc).
3. Briefly describe the problem and the target audience for whom technology is suitable (for example: what type of seeds, environments, farming practices).
4. Technology analysis in a simple way by answering 5 W (What, Why, Where, When, Whom) and 1 H(How).

   After the above-mentioned process, you should now develop the extension literature.
5. Draft the extension literature.

   Some tips to be followed is:
   - Be specific, simple.
   - Content should be culturally appropriate.
   - Logical, coherent.
   - Illustrated with images, figures.
   - Use active voice.
6. Pre-testing: Test content with the intended audience and consider their responses. By presenting ideas, communication content, and other extension materials to a representative sample of the target community, pre-testing enables you to verify and/or enhance the quality of your concepts, communications, and other materials. If pre-testing is less successful than anticipated, you can adjust concepts and materials or alter your approach. (Bell et al., 2015)

Guidelines for written materials

Brochures, factsheets, pamphlets, and bulletins are all types of printed materials used for various purposes. Although they may look similar at first glance, there are some key differences between them. Here's a brief overview of each type of printed material:

Brochure: A brochure is a printed document that typically has multiple folds and provides detailed information about a product, service, or organization. Brochures are often used for marketing and promotional purposes and can be distributed to potential customers or clients. They may include photos, graphics, and detailed descriptions of features and benefits.

Factsheet: A factsheet is a one-page document that provides concise information about a particular topic or issue. Factsheets are often used for educational or informational purposes and may be distributed at events or used in presentations. They typically include key facts and statistics, as well as visual aids such as charts or graphs.

Pamphlet: It can consist of single pager, or printed on both the sides. A pamphlet is a small, often pocket-sized booklet that provides information on a particular topic. Pamphlets may be used for educational or promotional purposes and are often distributed in large quantities. They may include photos, graphics, and short descriptions of key features or benefits.

Bulletin: A bulletin is a document that provides timely and relevant information on a particular topic. Bulletins are often used in organizations or communities to update members on news and events. They may include short articles, photos, and announcements.

Consider the following elements when designing extension literature:

The Basics: (4 W and 1 H)

- What do you intend to achieve?
- Who are your target viewers?
- What are the needs of your audience?
- Why do you believe this audience will find your message engaging?
- How do you anticipate your target audience to react to your message?

Start

- Create an outline for the extension literature.
- How do you wish to introduce and finish?
- How can it be broken down into manageable sections?
- What visual aids to use?
- When planning your extension content, solicit feedback from all relevant target groups.

Developing the extension literature such as brochure or bulletin

- Be clear
- Use relevant informative photos.
- Select fonts and images that are legible, sizable, and have good colour and contrast.
- Coloured materials are more enticing but ensure that a photocopied black-and-white version is also clear and readable in case individuals begin to copy the original.

Guidelines for Posters

A successful poster consists of

1. The most important facts only.
2. Use of easy-to-recognize pictures or symbols.
3. Strong colours and colour contrasts that are not gaudy.
4 A small amount of text that is easy to understand and is set in the right size font.

Posters can either get people's attention or give enough information for farmers to do something. Additionally, they should generate conversation. Posters should not provide an abundance of information. Use as little text as possible, emphasise photos, and make sure everything on your poster is necessary. If you wish to provide more details, create handouts. Posters are visual presentations, so show farmers what you want them to know.

1. Preparation and Layout
   - Determine the key message.
   - Write an action-oriented title, sketch your poster on graph paper to organise the content.
   - Information be presented from left to right and from top to bottom. Use arrows or numbers to guide the poster.
   - Keep it simple, cut down extra words.
   - Keep to 2 or 3 points.
   - Get feedback from others before you finalize the poster.
   - The size of posters can range from A3 Frame Size – 29.7 cm x 42 cm/ 11.75" x 16.5" to 150 cm x 90 cm / 60" x 40".
2. Text
   - Double-space all text.
   - Use short sentences and simple words.
   - Write concisely.
   - Avoid jargon, acronyms and/or unusual abbreviations.
3. Fonts
   - All text should be legible from 1.2 metres (4 feet) away.
   - The poster's title should be 80-150 points tall.
   - Subheadings should be a minimum 60 points.
   - Text should be 24 points.
   - Sans serif fonts read better. For example - Arial 11, Century Gothic, Franklin Gothic Medium, Lucida Sans

- The poster should use one font only.
- Bold, underline, or colour to emphasise.
- Capitalize just 1- or 2-word heads. All-caps text is hard to read.

4. Illustrations and images
    - Most graphics should be self-explanatory. Poster should feature such graphics. Keep captions concise.
    - Avoid visual distractions.
    - Make use of logos to demonstrate the institute and brand identity but ensure that they do not become a source of distraction.
5. Colour Use
    - Use only 2 to 3 colours.
    - Combine a dark background with lighter photos or use a light background with darker images.
    - Gray backgrounds highlight colour in photographs, while white backgrounds lessen their impact.

Activity

1. Create a poster on gender in agriculture.
2. Create a pamphlet on drone use in agriculture.

Notes

4

To Study About Presentation Skills Exercise

Learning outcomes

1. To make student aware about proper procedure of conducting oral and poster presentation and to develop their skill in delivering effective presentation.

What is presentation?

- Presentation is the process of demonstrating and conveying a topic's information to an audience or learner.
- Presentation may be customised for speaking to a group, addressing a meeting, or briefing a team.

Making oral presentations is nothing new for anyone, but there is still a lot of room for growth. An oral presentation is more than simply delivering the relevant technical information and using the perfect language. Public speaking is an address delivered at an anniversary or inauguration to encourage, congratulate, or amuse the audience. Speaking skills may be improved through practise and understanding of public speaking etiquette.

Written and spoken communication are different. The first is intended to be spoken while the second is intended to be read. However, they must both conform to the fundamental standards of excellent writing. Listeners have just one opportunity to hear you speak, and they cannot read when they are puzzled.

Two well-known methods for efficiently communicating your ideas are:

- First is to keep it short and simple. (KISS rule).
- Second, repeat crucial insights.

Oral message development

Face-to-face interactions include a speaker's personality, facial expressions, body language, voice, and speech delivery.

When preparing oral presentations, consider the following:

- Prepare adequate preparation
- Gather chronological information.

- Personalize notes.
- Logically order points.
- Do utilise visual assistance.
- Keep talks short.
- Be factual.
- Avoid emotions and theatrics.
- Show kindness.
- Use simple language.
- Exact concise clear.
- Make your discourse easy and entertaining; simplicity is beauty.
- You should not memorize the speech.
- Think about what you want to say.

The speaker wants to

- First, provide exact and accurate information.
- Second, verify audience comprehension of the message.
- Third, inspire audience acceptance and action.

Features of Good Speech

- Clarity - listeners must understand what you're saying
- In the style of a casual dialogue, successful speeches should be a great rapport between you and your audience.
- Concrete speech is needed.
- The speech should be specific, with easy-to-visualize information.
- Be concise -a person's attention span is limited; thus your speech should be brief and effective.
- Speech should be entertaining; incorporate quotes, stories, and humour. Good presentations are usually audience-focused and spontaneous.

Preparation work for speech

1. Determining the talk's setting- organisers, venue, occasion, time, other speakers' audiences/groups, age group, status, audience-relationship.
2. Preparing the text involves deciding what to say, gathering the necessary facts, and giving it structure through length, style, introduction, body of presentation and conclusion.
3. Oral communication involves - body language, personality, appearance, decent grooming, clothes and movements.

Some tips

- Comfortable posture.
- Eye contact with the audience.
- Do not move much.
- Natural facial expressions and body language.
- Address loudly.
- Make brief pauses when speaking.
- As health impacts your whole appearance, you should maintain a healthy body.
- No matter how much time and attention was spent on the creating content, the delivery is mainly responsible for the speech's success. Practice boosts confidence and time management.

4. In a welcome speech, the speaker tells the audience about the guest speakers and the event's host organisation. The speech opens with the welcome of the chief guest other important people and speakers invited. Do not have a long welcome note.

5. In farewell and send-off- are slightly emotional speeches. The content of the speech depends on the relation of the speaker and the person. If the occasion is send off to a higher position or over on a special assignment the tone is congratulatory the speech ends with good wishes for health, success, happiness, and further achievement.

6. Audience-focused

   Knowing your audience's age, background, and technical knowledge is vital for any kind of communication. What's simple for one may be hard for another. A colleague may know something a client or subordinate does not. People or groups may not know a company's technical jargon. It is the speaker's duty to make his audience comprehend and be interested in his subject. Presentation of data equals acceptance of views. You should consider the following:

   - How many people are expected?
   - What is the audience age range
   - Where do the listeners stand on issues of faith, politics, and economics?
   - How do you think the intended audience will react?

7. Incorporating Visual Aids

   Practice your presentation with the visuals. Most people remember considerably more of what they see than what they hear, thus it's important to practise synchronisation in visuals and speech, to become a good speaker and overcome stage fright.

Methods for using visual aids

Posters

Flip charts

Computer presentation

Collage

One of the following can be done to prepare for a presentation:

- The first is the hard sell approach, in which the speaker starts with a question or a comment that shows what the audience wants.
- Second, there is the "soft sell" method. In this method, the speaker introduces the company, product, or idea he wants to sell.

Presentation of research Paper

Research work can be presented in 2 ways:

- Oral presentation
- Poster presentation

Parts of Presentation

Opening

- Pose a question.
- Practice a lot.
- Draw the audience's attention.
- Briefly describe the issue and provide an overview of your subject.

The Body

- Present your essential ideas in a logical sequence, one by one.
- At the conclusion of each point, pause. Allow individuals time to take notes or think about what you're saying.
- Make it apparent when you go on to the next topic.

The Conclusion

- Give your viewers a concise overview of all you've discussed.
- Don't let the conversation go away. Make it clear that you have reached the conclusion of your presentation.
- Reiterate the important points.
- Restate the goal of your speech and state that you have succeeded:
- Say thanks to the audience and open the floor to questions.

Tips of presentation

- The voice should have an even tone.
- Audible to everyone.
- Reiteration of essential details is required.
- Don't rush anything; take a moment to pause.

Oral Presentation

There are two main formats for academic presentations:

- Classroom presentation
- Conference presentation

Oral Presentation

- The narrative should be shown using images or graphs.
- Use just a few slides for your Powerpoint presentation.
- The size of the text should be considered.
- The usage of animation effects should be restrained.
- A video or animation should be inserted.
- You should do more than simply read the slides.
- Numbers are often hard to understand.
- Think about putting information in graphs or charts.
- Clearly name your charts.
- The background shouldn't be distracting.
- Should have a light colour.
- Make it look like you know what you're doing.
- Don't panic.
- Practice presentation according to time provided.
- Practice makes you feel more confident.

For conference talks, for example, two rhetorical goals are recommended. Leave your audience with a clear idea of what you contributed and make them want to read your paper. Your presentation shouldn't replace your paper, but rather make people want to read it. It is often helpful to hint at information in the paper that can't be covered adequately in the presentation.

Research Poster Presentation

General tips

- Determine an essential concept
- Figure out the size of the poster you want to make.

- Make good use of your time.
- Draw it out! Sort the information into columns

Content of the Poster

The Title

- Putting the contents in order.
- Editing
- The text should go with the pictures.
- Check out what the experts on the poster say.
- Use active voice.

Picture and Text on Poster

- Self-explanatory pictures.
- Graphics should be able to be seen.
- Edit the content and visuals.
- Use 2-3 colours.
- Text on posters should be easy to read.
- Section headings should be bold

Poster background

- Softer colours would work better.
- Use a coloured background to unify your poster.
- Design a poster to answer one fundamental question.

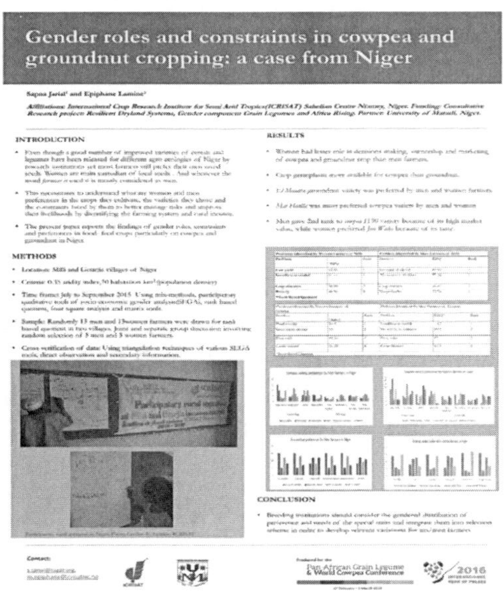

Image 1: Example of a poster presentation(Credit: Jarial, Sapna)

Posters are utilised in a variety of contexts, including displaying class work or presenting preliminary research in order to solicit comments and provide students with a chance to practice presenting. Also, businesses may use posters to demonstrate how they operate, therefore facilitating community input from anybody who may be a potential future employer or with whom you may choose to partner in the future and market or promoting yourself. Typically, poster presenting rooms are busy and loud. It is necessary to talk loudly. Therefore, the throat must be hydrated; carry water. Arrive early to the event and survey the area.

Consider the poster as a tool for presenting; hence, you should be professional and consider the same factors as when you consider presenting slides. You should construct a poster similarly to how you would a PowerPoint. The information on a poster is presented on a single large slide, and there is no gap between you and the audience. Additionally, interactive poster presentation is essential. The interactive element of a poster should allow you to engage in dialogue with the audience. Use questions to get audience attention.

Indicate the graphics that will aid in telling the story. Simply standing in place and telling a narrative. Bring business cards with your contact information for easy distribution.

Bring a notebook to record pertinent feedback information. Dress properly, you may discover a chance to meet a future employer. If you are presenting in a group, determine in advance who will present which piece, so that no one seems awkward in front of the group.

When presenting a poster in a busy room, you should avoid blocking the poster area. Maintain free space, maintain eye contact, and smile to make visitors feel welcome. For groups, stand on sides of the poster and maintain an open front. Presentation is a chance to convey yourself and your story in your own unique manner. Your potential employer may hire you based on your research and presentation.

Let us now examine the average seminar or conference presentations.

Most competent presenters spend an average of two minutes each slide, excluding title and outline slides; this is sufficient for a 20-minute presentation.

- One slide with the headings title/author/affiliation
- Give a summary of the difficulty you faced and the solution you arrived at. What is the one notion you want people to remember? This is the abstract of a presentation given orally. One slide.
- Provide outline in one slide.
- Give presentation structure.

- Background/motivation and problem statement-1-2 slides.
- Related work (0-1 slide)
- Cover briefly or leave out; tell people to read your paper.
- Methods- 1 slide- cover quickly in short talks refer people to your paper.
- Results 4-6 slides
- The most important results and insights should be shown. This is the most important part. Don't just quickly go over all the results; talk in depth about the ones that matter most. Don't just give numbers; show what they mean by explaining them. do not put up a big table of numbers.
- Summary -one slide.
- Future work- one slide. Optionally, list problems that this research raises.
- Backup slides- 3 slides. Optionally, have a few slides ready that aren't part of your talk but can be used to answer questions that come up.

Activity

- Discuss some dos and don'ts and body languages during Presentation?
- Prepare a presentation (poster or Oral)

Notes

5

To Study Micro Teaching Exercise

Learning outcome

The purpose is to enable the student to develop deeper knowledge, understanding, capabilities, and attitudes in the context of micro teaching skills.

Microteaching helps teachers worldwide improve easy teaching tasks. Microteaching promotes real-time learning experiences for novices and seniors. Presentation and reinforcement skills help inexperienced teachers acquire the art of teaching easily and thoroughly. The skill of teaching isn't only about transferring knowledge. It's a complex process that affects learning.

A teacher's quality is based on how much students understand. Classrooms cannot be used to master teaching basics. Microteaching is a teaching skill-learning method. Real-world teaching situations are used to hone skills and gain teaching expertise. Dwight W. Allen invented microteaching at Stanford in the 1960s.

Definition

"It is defined as a system of controlled practice that makes it possible to concentrate on specified teaching behavior and to practice teaching under controlled conditions". *- D.W. Allen & A.W. Eve (1968)*

- The experimental teacher teaches a small number of students.
- Each session's process is recorded.
- Record reviews, make corrections, and re-teach till desired outcome.

Micro-teaching is a teacher training technique that teaches teaching abilities.It requires the teacher trainee -

1. To teach one topic
2. Employing a specified teaching skill
3. For a short duration
4. To a few students

Steps for micro teaching

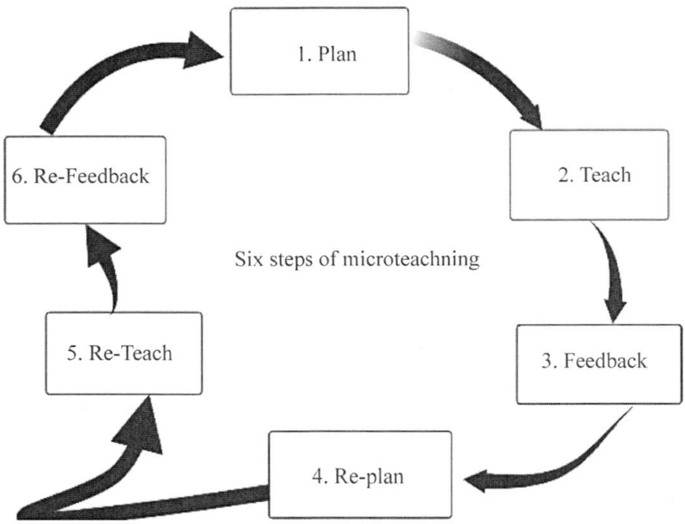

Image 1: Microteaching steps

1. Planning stage

It may focus on one concept. It must meet expressed or implicit goals. Plan consecutively and in writing. Methodology, not content, should be emphasized.

2. Teaching stage

Use skills in teaching-learning contexts. The teacher should adapt to class situations. To handle a class scenario well, he or she must have guts and confidence.

3. Feedback

This word means response to teacher trainees. It informs the student of the strong and weak points of the teacher. This improves the teacher's performance.

4. Re-plan

Teacher trainee re-plans lesson, capitalising on strengths and minimising weaknesses.

5. Re-teach

Depending on the subject, this may entail educating the same or a different group of students. This helps students avoid boredom.

6. Re-feedback

The most important aspect of micro-teaching for changing teacher-trainee behaviour in every skill practise is feedback.

Skills of Microteaching

Microteaching approaches are founded on the concept that teaching may be analysed and assessed using small teaching tasks/skills, which are the teacher's actions that enhance learning.

Lesson planning

It requires preparing a logically ordered micro-lesson. The material should be succinct, relevant, and fit the timeframe.

Presentation and explanation

This entails explaining topics clearly and correctly. Teacher excitement, a starting statement or theme phrase, good explanation, deliberate repetition, and finishing remarks or key messages with explanation summary.

Illustrating with examples

The teacher trainee should provide straightforward, relevant, and fascinating examples to help students grasp.

Reinforcement

This talent increases student engagement in instructional development. This skill requires favourable verbal and nonverbal signals.

Stimulus variation

A competent teacher must keep students' attention. Gestures, voice patterns, and interaction styles are effective components.

Probing questions

Allow other learners to ask structured inquiries and clarify misunderstandings. This talent involves redirection, refocusing, and critical awareness.

Classroom management

This ability requires giving clear directions, limiting unacceptable conduct, and addressing students by name.

Using audio-visual aids

This ability requires heightened audio-visual awareness. This ability requires adequate spacing, size, word and line spacing, and suitable words or phrases.

Merits and demerits

Microteaching has a pivotal role in all education training programmes and improves knowledge of the teaching process.

- The pre-and post-lesson plans had successfully demonstrated growth in teachers' knowledge on teaching.
- The "teach, critique, re-teach" model identified microteaching as a technique for personality development and confidence-building of professionals.
- Microteaching builds and refines skills.
- Improves communication, problem-solving, and critical thinking.
- Helps compare microteaching variations

Demerits

- Proper practise in teacher training programmes helps trainees get their first teaching experience and build knowledge, abilities, and attitudes. Microteaching in education training programmes is time-consuming. Neither all abilities are exercised nor are all trainees given a chance to re-plan and re-teach. The training becomes ineffective unless the teacher acquires the quality of effective student teaching.
- Large classes don't emphasise content, skill dependence, or administrative/ logistical issues.
- Microteaching is undervalued and underutilised. Microteaching's cheaper, simpler procedures would produce better instructors.

Activity

Prepare for a topic and present it for 15 minutes in your class.

Notes

6

To Visit the Village for Understanding the Problems Encountered by Villagers or Farmers

Learning Objectives

i) To raise awareness of a community problem and support its solution.

ii) Induce the emotional participation of the community at the local level and establish a positive psychological climate for new activities.

The various methods of data gathering involve the use of appropriate recording forms. These are called tools or instruments of data collection; they consist of an observation schedule or interview guide, an interview schedule, a questionnaire, etc.

There are two types of data

1. Primary
2. Secondary data

Primary Data- Researchers obtain primary data via interviews, questionnaires, trials, etc. Primary data is research data gathered directly from the original source via interviews, opinions, observations, events, or test findings (objects).

Secondary data- Secondary data is data obtained from primary sources and made accessible to researchers. It's data that has already been gathered. It's data that has already been gathered. For example, newspapers, books, websites, and journal articles.

For collecting the data, data collection instruments are used. A trustworthy instrument improves data quality. Reliable, consistent, and valid instruments are accurate. In other words, each time a reliable instrument is used, the same information should flow. Poorly worded survey instruments can affect study precision. As a researcher, you must carefully construct survey instruments to improve their quality and purpose.

Data collection techniques employ use of -

1. Questionnaire
2. Schedule

Questionnaire and schedule is a defined collection of questions for responders. Though similar, however from the technical point of view there is difference between the two.

Difference between Questionnaire and Schedules

S.No.	Questionnaire	Schedule
1.	Questionnaires are a non-contact way of collecting information; they are sent to informants with instructions, but no additional help from the sender.	A schedule is a straightforward technique for collecting information, usually filled out by a researcher or enumerator who can understand questions.
2.	Data gathering is inexpensive and economical because of questionnaire preparation and mailing. Comparatively cheaper and slower.	Data gathering is costly because of enumerators and their training. Money is spent on scheduling. This procedure is fast.
3.	Many individuals don't reply and return the form without addressing all questions. Non-response bias is frequently uncertain.	Non-response is minimal because enumerators who obtain all questions. Still, the bias and cheating by interviewers can be present.
4.	Who responds is unclear.	Identity of respondent is known.
5.	Many responders don't return the questionnaire, thus it's sluggish.	Enumerators gather information in a timely manner.
6.	No physical interaction is possible with surveys since responders return them via mail.	Formal link exists. Personal interaction is formed.
7.	This strategy requires literate, willing respondents.	Both educated and illiterate persons can be interviewed.
8.	Wider distribution.	Sending enumerator to wider area is costly.
9	Questionnaire performance depends on survey quality.	Depends on enumerators' honesty and skill.

This section discusses -

i) Survey instrument types,

ii) Interviewer role,

iii) Interview stages and

iv) Interview bias.

i) Survey instrument types

Survey research uses five main instruments (or approaches). They are:

a) Self-administered questionnaires,

b) Face to-face interviews,

c) Telephone interviews,

d) Internet survey, and
e) Structured observation.

a) Self- Administered questionnaire

A self-administered questionnaire is made up of questions that each person answers on their own. Self-administered questionnaires can be delivered or completed "on site," such as in a classroom, waiting area, or office. Respondents read and record answers.

Advantages	Disadvantages
1. Self-administered surveys are cheaper.	1. Low response rate
2. This tool allows respondents to carefully analyse questions.	2. Respondents may find questionnaires time-consuming and literacy.
3. No pressure to respond immediately, as in an interview.	3. In oral traditions-dominated cultures, questionnaires are less effective.
4. In a postal questionnaire, the researcher may send it to a vast geographical region and respondents can complete it when convenient and verify personal information, if required.	4. Reminders may increase response rates, but also increase data collecting time and expense.
5. Mailing surveys ensures anonymity and eliminates interviewer bias.	5. Incomplete and inaccurate questionnaires can pose a serious problem.
6. For a well-educated or interested target demographic, response rates are high.	6. Self-administered questionnaires restrict researchers' questioning. For example, questions requiring visual aids, open ended questions.

Techniques to improve rate of response

- Send follow up mail
- Notified in advance about the forthcoming survey
- Adding related picture to the questionnaire (help in some cases)

b) Face-to-face interview

An interview involves at least two people: an interviewer and a subject (the interviewee). Some circumstances allow group interviews.

Advantages	Disadvantages
1. Face-to-face interviews are optimal for survey data collection.	1. Face-to-face interviews provide benefits over self-administered surveys if conducted by skilled interviewers who earn respondents' respect.
2. Increases response and quality rate.	2. Expensive
3. Many responders who can't fill out a questionnaire on their own may answer to interview questions.	3. Interview bias can exist.
4. Doubt clarification is possible	4. The interviewer's look, voice tone, question phrasing, etc. might impact the response.

c) Telephone interviews

Telephone interviews are ones where the interviewer's voice is utilised. In many regions of the globe, phone calls are expensive, and the interview time is short. Open-ended inquiries are tricky. Interviewers notice only major interruptions (e.g. background noise) and respondent's voice tone (e.g. anger, flippancy or hesitancy).

d) Internet survey

Since the mid-1990s, online surveys have grown popular. Internet surveying has two major types: e-mails and web pages.

i) **E-mail surveys**: E-mail surveys are the cheapest survey method. Three kinds exist. First, as e-mail text-questions. This is the most basic method of e-mail survey in which respondents simply edit the mail message to indicate their response. Second, a formatted questionnaire sent as an e-mail attachment. Third, an interactive questionnaire, which can be sent as an attachment, answered, and returned by email.

ii) **Web Page based surveys**: The responder must visit a webpage (URL). This page contains a questionnaire. Pop-up questionnaires: When a person happens to visit the page, a questionnaire will pop up or the respondent will be asked if she or he wants to go to URL (web address) to answer a questionnaire. This form of sample recruitment relies on site visitors answering a questionnaire.

Advertising on other sites: Other sites may post surveys. This strategy may draw individuals from many places and generate many replies, but it's unclear what demographic such a sample represents.

Listserv, news group and chat group advertising: A researcher may advertise a questionnaire or invite participants from people on internet-based lists.

E-mail invitations: These may be sent to persons with addresses. Researchers may combine e-mail and web-based questionnaires. This instrument's flaws are similar to those of phone interviews.

e) Structured observation

A visually structured observation helps the observer concentrate on certain acts or qualities. Two visitors to a school are doing a structured observation if they are requested to count and record the number of computers, search for air conditioning, and measure the room's space in square feet. Participant observation uses it often.

ii) **Survey research interviewing** is a special kind of interviewing. The survey interview is a social relationship. The purpose is to get as much correct information as possible. Interviewing isn't simply question-and-answer. It provides much about the interview's social setting. Social roles, conventions, and expectations are involved.

The interview is a short-term, secondary social contact between strangers. Interviewer and the interviewee are social roles. Interviewer asks planned questions, and responder replies. It's different from regular chatter. Often, the interviewer or respondent's sex, ethnicity, or age affects their opinions and behaviour unrelated to the interview.

Many survey respondents are confused about their roles, and they frequently don't know what's expected of them. They replace another role, which affects their reactions. Some consider the interview a therapeutic session, others a bureaucratic form-filling exercise, etc. In a well-designed, professional survey, only half of respondents understand the questions as intended. Respondents reinterpreted questions to fit their specific circumstances or make it easier to respond. Interviewers play a key role. The interviewer needs collaboration, rapport, and neutrality. He invades the respondent's time and privacy for information that may not help the respondent. Investigators must decrease respondents' shame, anxiety, and suspicion so they may give information.

iii) Stages of an interview : There are three stages to an interview. The interview begins with an introduction and entry. The interviewer should explain why the selected responder can't be replaced. The interview is mostly asking questions and recording answers. The interviewer asks questions in sequence, without skipping or returning to questions until instructed. S/he maintains interest by moving at a comfortable pace and providing nondirective feedback. The interviewer should not summarise or paraphrase, as this causes knowledge loss or distorted replies. The interviewer should use probes effectively. A probe is a neutral inquiry to clarify uncertain answers, complete incomplete answers, or gather pertinent responses. Interviewers explore irrelevant or erroneous answers.

There are many types of probes: 3-5-second pause; non-verbal communication (e.g. tilt of the head, raised eyebrows, eye contact, etc.); question or response repeated, then pause, ask a neutral question (such as "any other reasons"?) In the last step, the interviewer thanks the respondent and departs. He then walks to a quiet spot to amend the questionnaire and capture additional facts. Date, time, and location of the interview; the respondent's attitude (serious, angry, laughing, etc.); the interviewer writes his own emotions on the side. "When asked about his money, the responder got uneasy."

iv) Interviewer bias : Interviewer bias includes six categories.

i) **Respondent errors**-Because of others presence, forgetfulness, shame, misunderstanding, or lying.

ii) **Unintentional errors** or **interviewer sloppiness**-incorrect respondent contacted, question misread, questions omitted, questions read in wrong sequence, wrong response recorded, misunderstanding of respondent.

iii) **Interviewer deception**- Purposeful responses, omissions, rewards, etc.

iv) **Interviewer's expectations influence**- about a respondent's answers based on the respondent's appearance, living situation, etc.;

v) **Interviewer failure to probe**- to probe properly; and

vi) **Interviewer's appearance influences responses**, tone, attitude, reactions to answers, or comments made outside of the interview schedule.

Other things may impact the interviewer. The interviewer's ethnicity, caste, gender, and social status may also impact replies.

Questionnaire Construction

All the data collection approaches (or devices) we've explored use a set of questions. This collection of questions is used for self-administered, postal, and online surveys, as well as face-to-face and phone interviews. This section discusses questionnaire types and types of questions, as well as how to create an effective questionnaire or interview schedule.

i) Types of Questionnaires are of three types- Structured, unstructured, and semi-structured questionnaires.

1. **Structured questionnaires** utilise closed questions that present a responder with a list of possibilities. It doesn't allow the responder much flexibility. These surveys are utilised when the responder is a high-profile person or professional who doesn't want a protracted interview.

2. **The unstructured questionnaire** solely open-ended questions, no response options for respondent.

3. **The semi-structured questionnaires** Most social research uses structured (closed) and unstructured (open) questions.

ii) Types of questions

a) **Structured (closed) and unstructured (open-ended) questions** Structured questions provide multiple-choice responses from which respondents pick. A structured observation collects data visually and is designed to guide the observer For example - How do you feel about, the new educational policy?

- Strongly Approve
- Approve
- Ambivalent
- Disapprove
- Strongly Disapprove
- Don't Know

b) **Unstructured (open-ended) questions-** do not predetermine the possible answers from which the respondent must choose. The following is an example. What do you feel are the major problems facing our education today? --- -- -- ------------------------ Instead of supplying answers, the questionnaire lets the responder write their own. Such questions enable respondents to communicate their deep sentiments and ideas about an issue.

Unstructured surveys generate data that is difficult to interpret statistically since answers must be categorised after collection.

c) Knowledge questions

Surveys show that most people cannot answer questions based on facts or knowledge. Because respondents do not want to appear ignorant, knowledge inquiries are perilous questions.

iii) Skip or contingency questions

Researchers avoid useless queries. Some questions are for specific respondents. A "contingency question' is a multi-part question. They are called screen or skip questions. Based on the first question's response, the respondent or interviewer is told to go on or skip questions.

Example: Were you a student of the Agricultural Social Sciences? (1) Yes (2) No (If No, please go to Question 2).

1a. If yes, which subject / discipline?

1b. How many years have you studied the subject?

iv) Guidelines for formulating a questionnaire

An excellent questionnaire is straightforward and usable. Design must reduce responder mistakes. Good survey questionnaires follow two rules: avoid uncertainty and consider the audience.

a) Introduction

Every questionnaire needs an introduction that describes the research. The introduction should provide questionnaire instructions and a confidentiality guarantee.

b) Questions' order

Most researchers start a questionnaire with straightforward questions, such as family size, employment, village, etc. It is vital not to make these questions so personal that respondents get defensive or end the interview. As a rule, the first questions are broader than subsequent ones.

c) Question format

Determine whether organised or unstructured questions are preferable for a specific goal. Is it better to use a numerical scale (0-9) or a verbal scale (very positive, moderately favourable, extremely unfavourable, somewhat unfavourable) in a structured inquiry?

d) Clarity

It is crucial to make sure that respondents will be able to understand the meaning of each question. In other words, the researcher must phrase questions in terms the respondents will understand, but he must be careful not to seem patronising to the respondents. This is often a difficult balance to achieve and is usually best gained through extensive pre-testing of alternative questions.

e) Cross-check questions

Ensure that responders comprehend each question. Include multiple questions on one subject to determine whether respondents answer them similarly. Such questions assess the questionnaire's validity and respondents' self-reports. "Random probe" is a specific cross-check query. Each respondent is asked to explain random questions. Various structured inquiries are answered by responders to accomplish this. This technique allows the researcher to exclude from the analysis phase any structured questions that were misinterpreted by many respondents.

f) Interviewer control

The questionnaire should elicit information and control the interviewer's job quality. The questionnaire should tell the interviewer what to do when a technique is questioned. The instruction, "Give the responder 2-3 minutes to consider", as some interviewees are impatient, therefore this may be essential.

g) Pre-test

In a pre-test, a small sample of respondents completes the planned questionnaire (anywhere from 10 to 100, depending on the size of the eventual sample). A pre-test may assist the researcher identify questionnaire issues. Questionnaires are revised accordingly.

v) What to avoid in questionnaire construction

a) **Avoid jargon, slang, and acronyms** while creating a questionnaire. Jargon and technical terms are difficult to understand, hence they should be avoided in the questionnaire. Slang is a kind of jargon within a sub-culture. Members of other sub-cultures may not understand it. Unless surveying a specialist group, avoid slang, jargon, and acronyms.

b) **Embrace clarity and precision**. Ambiguity plagues many questionnaire-makers. Researcher may assume without considering responses. For example, the question, "What is your income?" could mean weekly, monthly, or annual; family or personal; before taxes or after taxes; for the current year or last year; from salary or from all sources. The uncertainty creates inconsistent answers from various responders.

c) **Avoid emotional language and prestige bias**: Words may be connotative or denotative. Similarly, titles and positions bring prestige or status. Words having strong emotional implications and ties to high-status persons might affect how survey respondents hear and answer questions. Neutralize. Emotional language may lead readers to react to the words rather than the subject. Avoid identifying a statement with a prominent person or organisation. Instead, than addressing the subject, respondents may answer based on their sentiments.

d) **Avoid double-barrelled questions**: Make one subject per question. Double-barrelled inquiries combine two or more questions together. It ambiguizes a response. For example, if asked, "Are you provided a laptop and phone in the office?" the respondent who gets only laptop or the one who gets only phone will have problems in answering. The responder will not be sure about the response.

e) **Avoid leading questions**: Make respondents feel that all responses are valid. Keep them from knowing what the researcher wants. By its phrasing, a leading question urges the responder to pick one answer over another. For example, the question, "You don't spray pesticides, do you? leads respondents to state they do not use pesticides. Further, loaded questions can be stated to get either negative or positive answers.

f) **Avoid difficult inquiries**: Asking something few responders know irritates and creates low quality answers.

g) **Avoid false promises**: Do not start an inquiry with a promise and inquire about options. Respondents who disagree with the promise will be confused. For example, the question, "The library is open too many hours. Do you want it to open four hours later or close four hours earlier each day?" Those who embrace or reject the concept have no real option.

h) **Avoid asking plans:** Do not ask what individuals would do in hypothetical situations. Responses predict behaviour poorly. These questions are a waste of time since they're abstract and lack context.

i) **Avoid double negatives**: Double negatives are grammatically incorrect and misleading. For example, "I haven't got no job", logically means that the respondent does have a job.

j) **Avoid overlapping or unbalanced response categories**: Make response categories or choices mutually exclusive. "Mutually exclusive" means that response categories do not overlap.

Activity

Prepare 10 questions to understand the socio-economic condition of a farmer and pre-test in your nearby village.

Notes

7

To Study Organization and Functioning of District Rural Development Agency and Various Development Department at District Level

Learning Objectives

1. To evaluation of actual functioning of the District Rural Development Agency (DRDAs)
2. Guidelines for implementation of the various Rural Development Programmes.

Theory/Principle/Background of the topic

District Rural Development Agency (DRDA) is the district-level entity that oversees Ministry of Rural Development anti-poverty programmes. This organisation was initially founded to execute the Integrated Rural Development Programme (IRDP).

DRDAs were later given state and central government programmes. DRDA's job is to identify rural needs and implement relevant programmes has a technical, administrative, and financial role in scheme implementation. DRDA disburses cash for programmes and offers management and technical assistance.

MPs, MLAs, District Development Department employees, Bankers, NGOs, and underprivileged groups make up DRDA's governing board. District-level governance guides DRDA.

A separate DRDA Administration has handled administrative expenses since April 1999.This intends to make DRDAs more competent in handling anti-poverty programmes and an effective interface between the ministry and district level.

Rural development and poverty alleviation programmes are decentralised due to broad geographical regions, administrative needs, and the necessity to include local leaders and the community. Ministry of Rural Areas & Employment implements these programmes centrally. The Ministry is responsible for central funding, policy creation, general supervision, monitoring, and evaluation. Principal Secretary, Rural Development and Commissioner of Rural Development oversee rural development programmes at the state level. DRDAs implement district-level programmes.

Decentralization's context

With the adoption of the Constitutional mandate for decentralisation of powers, which gave these self-government institutions more power in planning for economic development and social justice, the idea came up to change the status of DRDAs and merge them with panchayats. Under the "devolution of functions" section, the 1st Round Table passed Resolution IV(i) to change the role of DRDAs. DRDAs were to be gradually combined with District Panchayats, and their technical knowledge was to be shared with all levels of Panchayats.

In this area, the progress that has been reported for all of India is far from satisfactory. There are some tough problems, and one of them is the "reported" opposition of some Union Government institutions to the idea of merging DRDAs with district Panchayats. The total merger seems to have only occurred in Kerala and Karnataka. This practise has a far longer history with the latter. In both states, the DRDAs operate as the Zilla Panchayats' cells for poverty alleviation and rural development programmes.

Reportedly, West Bengal has made similar actions, but they do not meet the criteria for a total merger.

DRDA Status in other state(s)

In 10 States/Union Territories, DRDAs remain distinct, with the sole relationship made with the Zila Parishads (ZP) being that the President of the ZP serves as the DRDA's Chairman. For example - Punjab, Himachal Pradesh, Uttar Pradesh, Uttarakhand, Madhya Pradesh, Bihar, Chhattisgarh, Odisha , Tripura, and Lakshadweep.

In Gujarat, the District Development Officer, who is also the Zila Panchayat's Chief Executive Officer, continues to preside over the DRDA. In nine States/ UTs (Assam, Goa, Haryana, Jharkhand, Manipur, Tamil Nadu, Dadra & Nagar Haveli, Daman & Diu, and Pondicherry), the Collector continues to serve as the Chair, or an alternative arrangement has been created. Andhra Pradesh has designated the President of the Zila Panchayat as the Chair of the DRDA and the Collector as the Executive Chairperson. Similarly, in Maharashtra, although the President of the Zila Panchayat serves as the Chair of the DRDA, the Zilla Panchayat's Chief Executive Officer is appointed as the Executive Chairman of the Management Committee.

Table 1: DRDA status in UT/State. Separate DRDAs; Collector is Executive Chairperson, although ZP President is Chairperson; several parallel entities across tiers.

S.No.	State or U.T	DRDA Status
1	Andhra Pradesh	Separate DRDAs; Collector is Executive Chairperson, although ZP President is Chairperson; several parallel entities across tiers.
2	Arunachal Pradesh	Separate DRDAs
3	Assam	DRDA distinct; ex-Director is Chair; various parallel organisations
4	Bihar	DRDA separate; ZP Chairman is Chairman of DRDA; many other parallel bodies;
5	Chhattisgarh	DRDA merged; the ZP Chairman is now the DRDA Chairman;
6	Goa	DRDAs separate;
7	Gujarat	District Development Officer chairs DRDA. Separate DRDA.
8	Haryana	ZP Chairman is DRDA Chairman and Dy Commissioner/ Collector is Executive Chairman. Separate DRDA.
9	Himachal Pradesh	DRDAs separate; ZP chairman is DRDA chairman; similar entities exist.
10	Jammu & Kashmir	DRDAs are split up; the DRDA administration is getting a facelift, which could be a good thing on its own; the Panchayat at the district level is called the District Planning and Development Board, which combines the ZP and DPC in a way.
11	Jharkhand	DRDAs separate; Dy Commissioner is Chairperson of DRDA
12	Karnataka	Since 1987, DRDA has been a part of ZP. It still has its own name, but only so that it can do business with the Ministry of Rural Development in New Delhi.
13	Kerala	The DRDAs have been merged with the ZP and now work as the ZP's cell for fighting poverty. Some parallel bodies (the Freshwater Fish Development Agency and the Brackishwater Fish Development Agency) still work as directed by the government of India.
14	Madhya Pradesh	DRDAs separate; ZP DRDA's Chairman is the Chairman. Many parallel bodies, especially those funded by multilateral and bilateral agencies and run by the State Government as parastatal entities.
15	Maharashtra	DRDA is a separate organisation. The ZP Chairman is the Chairman of the Governing Body of DRDA, but this is mostly for show. The CEO is the executive chairman of the Management Committee.
16	Manipur	DRDA is separate; Adhyaksha,Zilla Parishad, as the Head/ Chairman; and there are many parallel bodies that are mostly involved with the Union Government's development programmes and schemes.

17	Odisha	Programme Director -DRDA is designated as the Secretary-in-Chief of ZP, and the ZP President is the Chair of DRDA.
18	Punjab	DRDAs are split up; the ZP Chairman becomes the Chairman, and it is said that all Panchayats have access to the DRDA's expertise. There are other parallel bodies at the district and village levels (Village Water and Sanitation Committees, Village Education Committees) that have nothing to do with Panchayats.
19	Rajasthan	DRDA is said to have merged in 2003, but there are still para-statals and government schemes set up with money from outside sources.
20	Sikkim	DRDA is separate
21	Tamil Nadu	The District Collector is in charge of the DRDA, which is a separate organisation. There are many parallel bodies, such as village education committees, watershed associations, and village forests committees.
22	Tripura	DRDA is a separate organisation, and the President of ZP is the Chairman of DRDA.
23	Uttar Pradesh	DRDAs are separate. The District Magistrate is in charge, and the Chief Development Officer , who reports to the ZP President, is in charge of the DRDAs. There are also a few other parallel bodies.
24	Uttarakhand	DRDA is a separate body. The ZP President is in charge of it, and the other parallel bodies are Van Panchayats and Village Education Committees.
25	West Bengal	DRDA merged and changed its name to DRDC of ZP;
26	Andaman & Nicobar	DRDA is separate; Only one of the three ZP presidents is the head of DRDA.
27	Chandigarh	No DRDA
28	Dadra & Nagar Haveli	DRDA is separate
29	Daman & Diu	DRDA still works as a separate body and has not been merged with another.
30	Lakshadweep	DRDA is claimed to be integrated; the ZP president serves as the chair; DRDA provides assistance to the ZP; other parallel entities connect to central programmes such as the Sarva Shiksha Abhiyan.
31	Pondicherry	DRDA is distinct; attempts are made to connect other parallel organisations with Panchayats.

Activity

Identify the programmes undertaken by DRDA in your state/union territory.

Notes

8

To Visit NGOs to Learn from Their Experience in Rural Development

Learning outcomes

- To understand national and international NGOs and their functioning.
- NGO types and rules governing them.

Non-governmental organisations (NGOs), as the name suggests, are free from the government in its functioning. NGOs are groups or organisations of voluntary persons who do not represent any government action, and their members and associated persons manage all the functions. These numbers work as per the rules and regulations prescribed by the government, but they are independent of governmental control.

NGOs are not under the direct control of the government. Instead, NGOs work for human welfare and try to touch almost every aspect of life for the betterment and upliftment of the people.

How to register an NGO

In India, NGO can be registered as a

- Society
- Trust and
- Section 8 companies

It depends on the activity, as it is a voluntary group or institution that works for a social cause.

In common understanding, it is involved in carrying activities at a wide range only for the benefit of society at large. Civil society needs non-profits. It is the development arm of civil society. Voluntary organisations serve as social development accelerators worldwide. Voluntary work is part of India's culture and social heritage.

A Voluntary Organisation is a recognised Not for Profit organisation of individuals built on equality, charity, and volunteer labour to promote human growth and national building.

According to the Government of India Non-Governmental Organisation Partnership system, Voluntary organisations (VOs) / Non-Governmental Organisations (NGOs) consist of Organisations engaged in public service based on ethical, cultural, social, economic, political, religious, spiritual, philanthropic or scientific and technological considerations. Following are the organisations that are included in the above category.

- Community-based organisations (CBOs);
- Non-governmental development organisations (NGDOs);
- Charitable organisations;
- Support organisations;
- Networks or federations
- Professional membership associations.

In other words, voluntary organisations are legally constituted organisations operating independently from the government and are generally considered to be "non-state, non-profit oriented groups who pursue purposes of public interest".

2. A brief history of non-profits

International NGOs date back to 1839. In ancient Greece, various VOs ranged from clubs of the affluent to religious or professional associations. In the Middle Ages, Guilds helped the government with some administrative tasks. It is thought that by 1914, there were 1083 non-profit organisations worldwide working on issues like slavery, women's right to vote, and disarmament. When the United Nations Organization was made in 1945, the number of volunteer groups worldwide grew significantly.

3. What is behind the rise in non-profits?

Recession, the end of the cold war, increased privatisation demands, and globalisation in the 20th century are the fundamental causes.

National Scenario: India has a long tradition of volunteerism focused on "Dhana" (giving) and "Seva" (service). Rabindranath Tagore's Santiniketan efforts revealed how to integrate education and culture before independence. Mahatma Gandhi's Wardha experiment illustrated how village industries might help the needy. In the second part of the 19th century, nationalist awareness developed across India and Self Help gained popularity. The early examples of such attempts are:

- Family in Need Society (1858)
- Prathana Samaj (1864)

- Satya Shodhan Samaj (1873)
- Arya Samaj (1875)
- National Council for Women in India (1875)
- The Indian Conference (1887)

NGOs emerged in India soon after independence. Many followers of Gandhiji established voluntary agencies to work closely with governmental programs on social and economic issues. These organisations coordinated handicrafts, village industries, and rural development projects. Many individuals realised in 1960 that government programmes were not helping India's underprivileged. So, the grassroots groups worked on a small scale with few resources and no coordination.

Since independence, the Indian government has made little effort to define or acknowledge the function of voluntary agencies. In 1980, with the Sixth Five Year Plan (1980-1985), the government recognised new development areas for NGOs. Among them:

- Optimal use of renewable energy, including forests, via block-level renewable energy organisations
- Welfare, health, nutrition, education, and community projects in the field
- Health
- Water management and soil conservation
- Social welfare programs for weaker sections
- Implementation of minimum needs program
- Disaster preparedness and management (i.e. for floods, cyclones, etc.)
- Promotion of ecology and tribal development,
- Environmental protection and education.

The voluntary sector soon became more specialised, and the voluntary movement was fragmented into three major groups:

Voluntary Organisations Type Modes of Operating

Example

1. Non-profits in Five-Year Plans

1. Volunteer groups should be more significant in making communities as self-sufficient as possible.
2. Voluntary Organisations were supposed to demonstrate how to employ village and indigenous resources and leverage human resources, rural skills, and local knowledge for their growth.

3. The significance of NGOs in rural evaluation for putting up low-cost development plans and incorporating the rural population is emphasised.
4. The Indian Planning Commission promotes a national network of NGOs.
5. It advocates NGO involvement in public-private partnership development.

2. Present status of Voluntary Organisations

UNDP's Human Development Index for 2021/22 puts India at 132 out of 191 countries and territories. This index tracks the global decline in human development. The 2018 Global Hunger Index shows that 38.4% of children under five in India are stunted. India is ranked 107th out of 121 countries, with enough data to be included in the 2022 Global Hunger Index. India has a very high level of hunger, with a score of 29.1. In such a setting, civil society raises people's concerns and protects minorities.

3. India's NGOs' legal status

Five procedures exist for incorporating a non-profit in India:

1. The Societies Registration Act of 1860
2. The Indian Trusts Act of 1882
3. The Co-operative Societies Act of 1904
4. The Trade Union Act of 1926, and
5. Section 25 of the Indian Companies Act, 1956.

4. National Policy on Voluntary Organisations 2007

This policy recognises India's NGOs. It signifies the government's resolve to support, facilitate, and empower an autonomous, innovative, and booming volunteer sector that can contribute to India's social, cultural, and economic progress. In addition, this policy acknowledges the necessity for local, provincial, and national coordination between the government and the non-profit sector.

Objectives of the Policy

- VOs need an enabling environment for autonomy and efficiency.
- Enabling VOs to mobilise national and international financial resources within the legal framework.
- Find out what systems help with public-private partnerships.
- Promoting good governance among VOs.

The policy talks about how important it is for the government to encourage all central and state government agencies to add modules to their pre-service and ongoing training on working well with non-profits. It acknowledges the challenges encountered by the non-profit sector in evaluating government programmes and provides solutions.

Activity

Visit a nearby NGO in your locality and create a case study.

Notes

9

To Study Participatory Rural Appraisal Techniques with Their Application in Village Development

Participatory Rural Appraisal (PRA) is a set of tools and techniques used with households to gather and analyse information on community resources, problems, potential and needs.

Why is PRA used?

Look at how a village or community is now and what it could be.

Look at problems and what's causing them.

Help families find ways to deal with problems and take advantage of opportunities.

How is PRA Conducted?

PRA is done with a group of households from a Village or Hamlet who work with "Facilitators. "Facilitators help groups of families. It is not giving lessons or lectures, rather facilitators work with and listen to families to find out what they need.

Materials and Equipment

Locally available materials, for example liker small pieces of coloured paper ('cards') are useful— on them, photos or labels may symbolise objects.

Paper and pens—these are used to record the results of exercises. A4 folders, A4 paper and coloured pens are needed for Facilitators to record the results.

Chalks, locally available stones, pebbles, food materials, spices etc ., can also be useful for making drawings on the ground.

Principles of PRA

PRA's core ideas have changed over time. It's interesting that new ideas keep being added to the list. The distinguishing characteristic of these principles is that they are derived from practise and knowledge of what works and what does not. Chambers (1997) has listed the following principles.

1. **Listening and Learning**. PRA is founded on the principles of listening and learning via interactive participation and progressive learning. The local populace has their own information, experience, history and culture, perspectives and beliefs, and tastes. Listening to locals helps expose their "worldview," which might otherwise stay hidden and unrevealed. The stronger the engagement with otherwise untapped ability of a listener as opposed to a speaker, the more learning is attained. This learning might improve incrementally. A learner's mental and physical setting must be optimal for learning to occur effectively. If a student intends to acquire knowledge, he or she must be psychologically prepared to listen, study, and show respect for those from whom such knowledge may be obtained.
2. **Offsetting Biases.** PRA is meant to counteract the biases that often come with "rushed" evaluations by professionals who want to get quick results. Professionals prefer to evaluate samples at opportune times when the weather is favourable, consult would-be-beneficiaries (mainly males) who are typically more affluent and focus on those elements they deem significant. To counteract these prejudices, PRA emphasises comfortable listening and learning, seeks the engagement of individuals who are generally less fortunate, visits rural and interior regions, and visits local communities at their leisure. This process should try to include people who would never have a chance to speak up otherwise. At first, there may be some reluctance, but the locals start to enjoy it as they learn new things about themselves, such as that the poorest people, women, and disadvantaged groups in remote areas, etc. Still, it's important for the team not to make any value judgments about other people.
3. **Utilization of Precious Community Time.** PRA uses valuable community time to learn from locals. It's crucial to maintain track of local community members' time and learn as much as possible. This also means members should engage at their convenience. Learning should be concentrated to maximise this time.
4. **Seeking Diversity**. PRA learns from many situations and actors. It seeks greater diversity in findings and data rather than representativeness. Diverse events, processes, and forces assist grasp challenges from many viewpoints. Greater variety improves "reality" analysis.
5. **Triangulation /Cross checking**. Triangulation improves data credibility. It will be done by changing the people on the team, where the information comes from, and the methods used. Each activity or event needs to be

looked at from different points of view and studied using different methods. Cross-checking is an important PRA principle for reducing mistakes and making changes in the middle of a project. Since information can come from many different places, it is important to test it in different ways to make sure it is correct and reliable. Changing methodologies, places, time, groups, teams, etc. may help assess PRA data authenticity.

6. **Optimal ignorance and appropriate imprecision**. The notion of optimum ignorance helps PRA facilitators reduce cost and time. It implies understanding what's worth knowing and enough to fulfil the purpose, but not more. Seek suitable imprecision or avoid unnecessary precision. These PRA principles assist make learning iterative, facilitate participation sessions, and make processes dynamic, inventive, and informative.
7. **Multi-disciplinary Team**. The PRA team must include scientists from all relevant fields. Female scientists are also required so rural women may participate in the evaluation. The team should choose a facilitator. Identify a Process or Content Recorder. While interacting with locals, he/she should first build rapport. This task could be assigned to one person.

PRA shifts focus from dominating to empowering, closed to open, individual to group, verbal to visual, measuring to comparing, rating and scoring, and reserve to rapport and shifts from irritation to enjoyment.

PRA Tools

- Transect Walk
- Venn Diagram
- Resource Mapping
- Seasonal Calendar
- Social Mapping
- Matrix Scoring

Transect Walk

Transect walks and guided field walks -- The researcher and key informants go around regions of interest to watch, listen, identify zones or situations, and ask questions to find issues and answers. This strategy teaches outsiders about terrain, soils, land usage, forests, watersheds, and community assets.

Process

- Women, men, young, and elderly should form 2 or 4 groups.
- Either participants may take various team members on separate transect walks to illustrate their most important places, or each group can be responsible for a different issue while walking together.

- One group may concentrate on soils, land use, and agriculture, another on trees, plants, and water, and a third on infrastructure, housing, and services. The groups exchange their walk data to create transect diagrams.
- Choose a straight route across the region using the Village Resources Map and participant input. The selected line should include as many physical zones, plant types, land-use regions, and community parts as feasible.
- Start at the area's highest point. Depending on the area to be traversed and topography, a transect may be done on foot, animal, cart, or vehicle. Slower settings offer more observation.
- Materials- use Notebooks, pens, flip chart paper and markers.
- Ask questions about each zone while transecting. Notes are taken along the transect. During the transect walk, interview women and men you meet. Discuss the team's identified major concerns and ask if there are more. When performing a transect, individuals are more likely to address delicate matters like land ownership patterns. Allow ample time for the transect. It may take many hours.

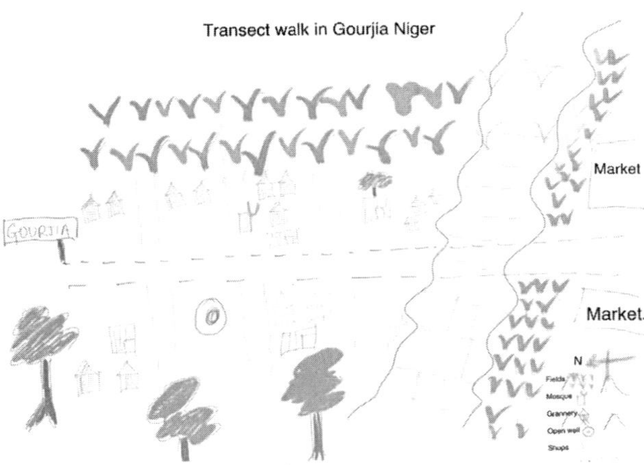

Image 1: Transect Walk in Transect **by Men**. Date: 3.07.2015. Location: Gourjia, Niger. Participants: Malicki Dan Abdou, Issofou Faroukar,Moctar Iro,Boulama Maman,Ali Sani Facilitator: Epiphane M Lamine, Assistant: Indo Sani, Recorder: Kimba Noufou

(Source: Jarial, Sapna)

Resource mapping

Resource maps show the location and status of a community's natural resources.

Purpose

Village Resources Map lets us understand a community's resources. Not cartographic correctness, but local views of resources is more important. Participants should identify what's essential to them.

Maps may include

- Water bodies, irrigation supplies, and drainage systems.
- Crop growth, patterns, production, etc.
- Terrain, topography, slopes.
- Forest, plant species
- Land tenure, ownership, and usage.
- Grazing areas
- Shops, markets
- Health clinics, schools, and religious facilities
- special use places (bus stops, cemeteries, shrines).

Process

Plan a community gathering. Make sure women and men may attend and invite all socioeconomic classes.

- The Village Resources Map is an excellent first tool since it promotes interaction between community people and team.
- Clear a huge open area. Place a rock or leaf to indicate a focal landmark. Participants mark essential village features on the map. Participants shouldn't be stopped until they stop sketching, at which point important questions may be asked.

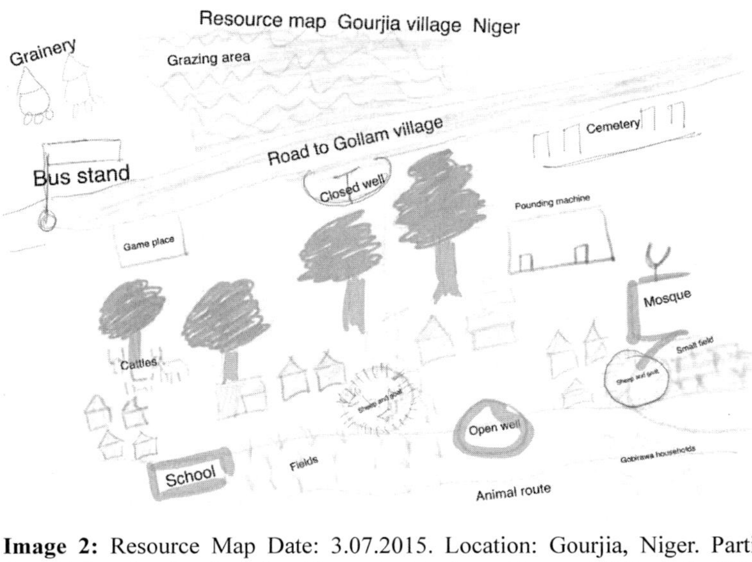

Image 2: Resource Map Date: 3.07.2015. Location: Gourjia, Niger. Participants: Malicki Dan Abdou, Issofou Faroukar,Moctar Iro,Boulama Maman,Ali Sani Facilitator: Epiphane M Lamine, Assistant: Indo Sani, Recorder: Kimba Noufou (Source: Jarial, Sapna)

- Facilitators should invite participants to describe and discuss the finished map. Clarify any confusion.
- Finally, the facilitator may invite participants to sketch something they would want to see in their community that isn't on the map. This allows for basic planning ideas and promotes early participation.
- Materials- Sticks, stones, leaves, sawdust, flour, manure, or local materials. Use flip chart paper and markers.
- The mapping activity orients team members to the community's geographical characteristics and vital resources. Make sure the finished map contains North, South, East, and West arrows, and village boundaries. Thanks, others, for participation.

Venn diagrams

Venn diagrams -- People, villages, or institutions are represented by overlapping circles, and inputs and outputs are shown by lines. Largely distant circles mean little contact or cooperation; only loose contacts exist if circles are touching circles and some co-operation. Overlapping circles means close co-operation.

Steps

- Tell participants the exercise's objective.
- Ask them to mention institutions, people, etc., based on the exercise's goals.
- Create little cards. Non-readers require pictures.
- Ask participants to rank one variable, such as institutions value, on a card.
- Ask whether they agree with the sequence after the cards are sorted. If they're interested, encourage modifications. Ask whether they agree with the sequence after the cards are
- Sorted. If they're interested, encourage modifications, if required.
- Ask them to allocate paper circles of varying sizes to institutions or persons such that the larger the circle, the higher they rate on that variable.
- Paste institutions or names on circles. You may write institutions or people on the circles.
- Circle the community on the ground. Ask them to put the circles so those with great accessibility are near together and those with poor accessibility are far from the community circle.
- After placing the cards, ask whether they agree.

- If institutions/people interact or cooperate closely, they might be placed with an overlap. The degree of overlap indicates the cards in such a manner. Note down the points of discussion and explanation.
- Create a copy. Record the village, participants, date, legends, circle size, and distance.
- Thanks to everyone who participated. Triangulate the diagram and primary finding to verify accuracy.

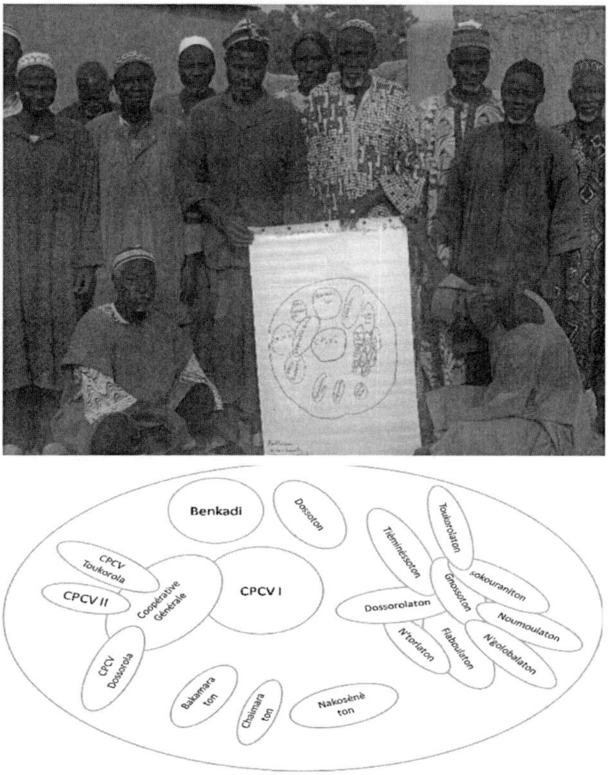

Image 3: Venn Diagram by Men in Mali. (Source: Jarial, Sapna, 2015)

Social Mapping

Social mapping comprises sketching/drawing residences, social amenities, and infrastructure (temple, shops, rice mills, school, pharmacy, trails and roads, water pumps, irrigation, and leisure facilities) in a community.

- Developing a comprehensive understanding of the physical and social aspects of village life.
- Collecting demographic and other required information household wise.

- Providing a forum of discussion in which to unravel the various aspects of social life.
- Serving as a guiding instrument during the process of planning interventions.

Steps

- Consult locals about the exercise's location.
- Tell participants the exercise's objective. Start by sketching the nearby landmarks. Let them utilise local and other items creatively.
- Be vigilant. Carefully follow the conversations. Detail your notes.
- Slow down. Don't contribute. Try handing them the stick, or actively involving them. Let them take charge. Show them your faith.
- Keep track of who's participating, which groups they belong to, and who's left out. Engage those left out.
- Wait to add or explain if necessary. Don't interrupt. For example: Ask "what about...," "what does this sign mean?"
- After mapping, ask individuals to find their houses in the map.
- Map interview If you're unclear about anything, ask for clarification.
- Create a large-scale map. Don't botch up details.
- Triangulate local data.

Seasonal Calendars

Seasonal Calendars enable us study yearly variations in livelihood systems. They overcome temporal biases by showing what occurs in various seasons. Calendars may be used to analyse how much individuals work at various times of year or how their wages fluctuate. It shows the seasonality of food, feed, crops, livestock markets, and water availability.

Process

- The focus groups of women and men created earlier, ask about people's yearly activities.
- Each group needs a vast area. Calendars may be created on huge paper or in the sand or mud using stones or leaves.
- Across the cleared area, draw a line (or paper). Explain that the line symbolises a year and inquire how people divide it (months, seasons, etc.). The participant-friendly scale should be used. Mark seasonal divides along the line's top.

- Rainfall patterns are a good place to start the calendar. Ask participants to place stones beneath calendar months to depict rainfall (where more stones equal more rain).
- Once the rainfall calendar is complete, you may ask them to build another calendar depicting their agricultural labour (putting more stones over the time periods of high labour intensity). Make sure the labour calendar aligns with the rainfall calendar.
- Repeat until all seasonal difficulties are handled. Include food, water, income, and expense schedules. Ask participants to label each calendar with a symbol or sign. Ask participants to detail food and money sources, etc.

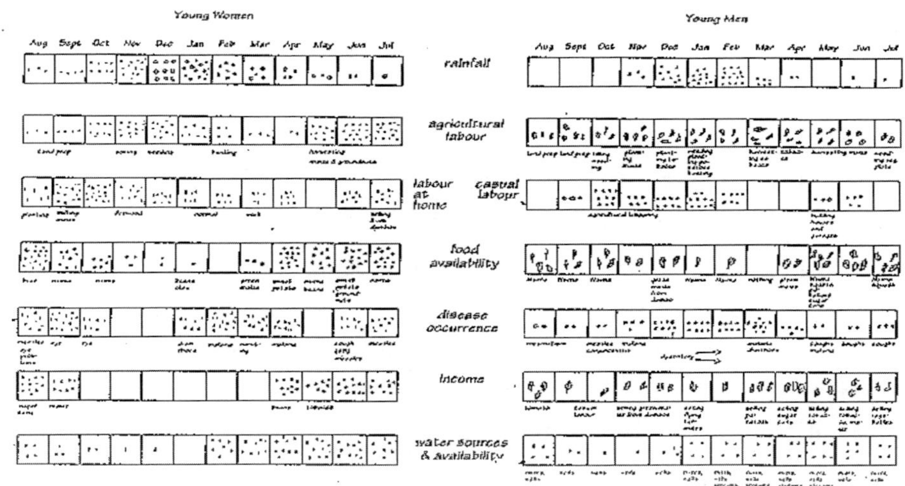

Image 4: Seasonal Calendar (Source, SEGA FAO, 2001)

Matrix Ranking and Scoring

Objective is to rank and prioritize specified things (e.g. shrubs and plants), objects, issues, solutions, project ideas; to create reasons why individuals prefer one item over another. to compare groups' perceptions and Strength , Weakness, Opportunity, and Threat (SWOT) analysis.

Matrix Ranking is used to list items to be compared on one line and criteria on the other line to rank choices from most important to least important(i.e. 1st, 2nd,3rd, 4th etc.) . In this case frequency of the items valued as the 1st choice helps to make up a final decision. Example-

Matrix Ranking of Most Efficient Mode of Transport

Element	Judging criteria				
	Cost	Frequency	Availability	Energy	Time
Walk	1	1	1	5	5
Car	5	4	5	1	1
Bus	4	5	4	2	2
Bicycle	2	2	3	4	3
Cart	3	3	2	3	4

Table 1: Matrix Ranking.

Matrix Scoring

- Put a score next to each of the listed criteria before making a choice.
- Many things could be compared to each other.

Matrix Scoring Technique

Crop Preference -Matrix Pairing Tool

1. Choose a person or a group
2. Select or invite others to choose a class of significant, well-known things such as crop variety.
3. Ask for the top ones. This list might have 2-7 items. Four, five, or six works best.
4. What's good about each item? What's bad?
5. List criteria. Turn negative criteria (e.g. pest-vulnerable) into positive ones (e.g. not pest-vulnerable)
6. Make a matrix using the objects across the top and the criteria down the side.
7. Enquire which meets each requirement best?

   With six objects the following sequence works well: - Which is best? Which is next best? Which is worst? Of the remaining which is better?
8. Rank directly in the matrix.
9. If you could only choose one, ask. Which would you use? Which next? Which next?

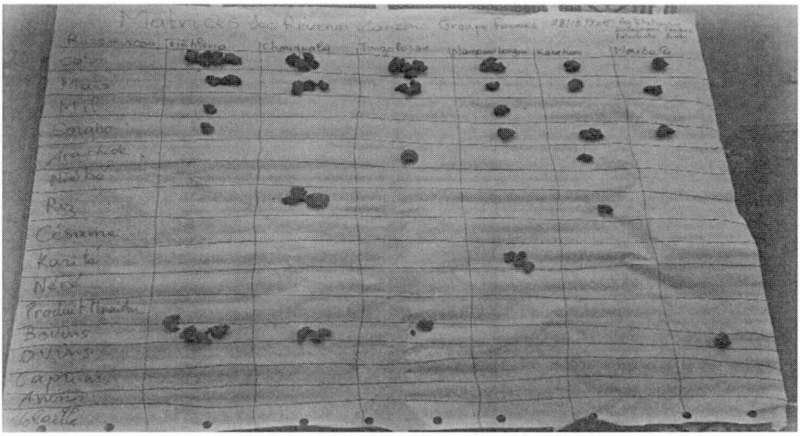

Image 3: Matrix Scoring in Mali (Credit: Jarial, Sapna ,2015)

Activity

Find out the crop preferences in the nearby village using matrix scoring.

Notes

10

To Visit the Community Radio Television Studio for Understanding the Process of Programme Production

Learning outcomes

- To understand the functioning of community radio in dissemination of from technology.

Introduction

Radio is the oldest type of media, and via radio the information revolution saw the green revolution. Farmers listen to the radio for weather, market price, and pestilence updates. Government began to encourage citizens by establishing radio rural forums (RRF). The primary purpose of a rural radio forum is to discuss and act. Here, a group of twenty individuals debate cultivation-related concerns and provide questions for the subsequent Radio rural forum. This allowed radio forum to get widespread popularity.

- Technology is always evolving. The cost of transistors has decreased, and there has been a move to individual listening from group hearing.

Types of radio

1. Public radio
2. Private radio
3. Community radio

1. The objective of government-owned and -operated public radio is to serve the public in the manner deemed best by the government.
2. Private radio operated by individuals or businesses with the goal of stationing a large number of private stations in our nation
3. Community radio is distinct in that it is financed by the state, individual corporations, and non-profit organisations or cooperatives whose members are programme listeners.

Community radio is a third radio paradigm, after commercial and public Community radio serves geographic and interest groups. They transmit popular, local material that commercial or mass-media broadcasters neglect. If public radio is an elephant, then community radio is like a mosquito. The communities they serve run, own, and influence community radio. They are generally nonprofit. Individuals, organizations, and communities may tell their own tales, exchange experiences, and become media makers and contributors.

Community radio stations should concentrate on getting the community talking, not only on radio. Community radio emphasizes social issues over radio. Mainstream stations are seen as catering to commercial interests or presenters' personalities.

- The Community Radio Policy announced by Government of India in December 2002 and later liberalized in 2006 as a result non-profit NGOs, autonomous bodies, and educational institutes etc. were eligible to establish a community radio station (CRS).
- The organization should be a registered body and should not have any political ties.
- The policy enables 100-watt CRS transmitters (250 watts in special cases).
- Permissible maximum antenna height is 30m.
- It should cover 5-10 kilometers.

CRS - Criteria for Site Selection

Government provides subsidies for the establishment of CRS. The money, however, must be spend before. One should consider the financial resources, terrain profile, aavailability and reliability of power supply, manmade and ambient noise, vventilation and freedom from dust. In addition, assess the lightening and water damage, availability of space to erect towers, accessibility to the participants , acceptability by the community and security considerations. Earlier many forms were to be filled in for applying for a CRS, which was a complicated process, now, however, the government has made this process easy by decreasing the number of forms.

Inadmissible criteria for CRS

- No CRS can be made operational without getting all the licenses from Ministry of Information and Broadcasting(I&B) and WPC (Wireless Planning and Coordination, Ministry of Information and Communication) and signing of a Grant of Permission Agreement (GOPA) by Ministry of I&B.

- The transmitter may only operate on the FM band frequency assigned by WPC(87 to 108 MHz). No other frequency is acceptable. Therefore, the use of a transmitter, such as that of a cordless telephone, is prohibited.
- There are rules governing the standardisation of transmitters in terms of frequency stability, frequency drift, production of harmonics, etc. No one may manufacture and operate his own transmitter unless these have been examined and a licence for manufacturing transmitters has been acquired from the government.
- The tower height has to be a minimum of 15 meter to protect biological hazards. The tower must be suitably designed so that it withstands the wind, storm and rain and does not fall and causes injury.
- Programme must follow the All India Radio (AIR) Code.

It is a broadcasting organisation that provides social, economic, and cultural communication assistance to the community. It runs in a village district and is owned and administered by the community on a non-profit basis. The notion of community radio focuses primarily on the word "community," which refers to a group of people who have something in common. This community may be physical or virtual, and the community radio -radius is 15 to 20 kilometres. Community of interest radio is another kind of virtual community radio; it focuses mostly on low-power, low-cost wireless broadcasting sessions. It acts as a catalyst for social development, caters to the interests and cultural demands of marginal, isolated, illiterate populations, and provides a supporting medium for impoverished women. It emphasises women's rights on political and social concerns and maintains community solidarity through natural calamities such as flood and drought. The listener has a strong feeling of affiliation with the programme, which is notably unique from radio stations.

Operationalization of Community Radio

It has 3 parts.

- First studio for live streaming or programme recording for later broadcasts
- Second transmitter for voice signal conversion to high frequency for transmission
- Third tower with antenna mounted atop a high pole or tower that is used to convey signal across wide distances, it also manages and operates the equipment daily, with the assistance of volunteers for the most part. For hardware maintenance content production, school instructors, students, and other diverse groups may be formed and given the task.

Running a CRS

- One should establish regulations for managing volunteer, fundraising, liaison with local administration for program production.
- One should prepare annual financial statements, attend meeting with anyone who supports connecting with programme staff to guide and encourage their work.
- Community members are the target audience, and suppliers of equipment, instruct the community members on how to operate and maintain the equipment.
- at government- local administration authority, the department which will grant permission for the running the program either through AIR channels or through ownership of radio transmitter.
- Producers and staff are comprised of experts in production techniques; they generate ideas and develop programme aids with input from other sources.
- Resource mobilizers are those who give funding, such as a state corporation or non-governmental organisation; nonetheless, the ultimate source of sustainability must come from the community. Community radio focuses on farm wives, rural youth, jobless rural youth, and traditional, health-related, and other topics. Children may contribute to singing, physician interviews, and environmental influence on social soil, water, and air pollution. Labour concerns relevant to policymakers and others might be used as material.

Challenges before CR initiative

- Getting license and regulatory issues
- Cost effective technical solutions
- Sustainability of the CRS
- Content sharing and capacity building of the local people.

Example - The Indian government has been supporting content sharing, particularly radio programmes, best practises, and case studies, to enable the effective use of existing resources.

Edaa is a web-based service that uploads radio station material. Listening to Bhojpuri or Tamil from villages that don't even exist on Google Maps is such an interesting platform that it is included in the ministry's news release on the future of Community Radio. Edaa is the largest community-produced audio archive in South Asia and offers over 2,900 radio programmes in 28 distinct

South Asian languages categorised by 33 theme groups. The Community Media Manch Platform is another online arena that facilitates learning and information sharing amongst Community Radio stations in India. This portal enables community radio stations and community media participants to engage in partnerships, information exchange, and webinars to share their expertise.

Community radio will be able to do the following via its regular operations:

- Providing a venue for community growth;
- Encourage participative community development;
- Promote the active participation of disadvantaged populations such as women and youth;
- Increase information exchange among the community;
- Promote creativity in community development;
- Improve the free flow of accurate and balanced information to the community and within it;
- Providing a venue for local cultural expression;
- And improve people's access to information in local languages

Presently people are creating podcasts, which are like radio shows. Following is an example for creating a podcast relevant for listeners.

Bonus Material

Steps to Start a Podcast: Complete Step-by-Step Guide

1. Develop and define a podcast theme.
2. Select your podcast format
3. Install podcast recording equipment.
4. Select your podcast software
5. Record.
6. Edit audio recording.
7. Create podcast's cover sketch.
8. Set up a podcast in Google podcast, Spotify
9. Launch and expand your show.

1. Develop a podcast concept

- The right podcast theme will attract and maintain listeners.
- Follow these techniques

Define your goals

- Why do you want to start a podcast? It's important to know your why.? Some good reasons for starting a podcast are:

- to get people interested in a business
- share things important

Some podcasters quickly become well-known, but most have a small group of listeners that they build up over time.

Pick a theme or topic for your podcast

- You should be excited to talk about and learn more about your podcast topic on a regular basis. The less likely you are to get podfade, the more you enjoy your topic.
- Once you have some ideas, do some market research to survey the competition in your genre.
- Try coming up with ten-episode ideas to test out your own podcast idea. If you have trouble coming up with ideas, you might want to change the subject.

Make it clear

- When you focus on a niche, you attract people who are more interested in what you have to say. Podcasts that appeal to a wide range of people may reach fewer people in the end.
- People are also more likely to tell their friends about a niche podcast than a general podcast.
- Focus on a sub-category or sub-demographic to make your podcast stand out.
- Focus on what your podcast does differently. What can you say about the topic of your podcast that no one else can?
- Examples- Starting a Food Producer Company (FPOs)

Name your podcast

- The name of your podcast is an important part of your brand. It gives people a good first impression of your podcast and sets the tone for it.
- The names of podcasts should be clear, direct, and easy to remember, like your business name or brand. Once you have an idea for a title, you can use a podcast name checker to see if the social accounts and domains are free. You shouldn't also use the name of someone else's podcast.

2. Select your podcast format

- Your podcast's format should fit it and give the impression that it can be sustainable.

- Think about what makes you feel good.
- Do you love getting to know your podcast guests, or do you feel best when you're on your own?
- If you choose to have cohosts, try to keep the group small. When there are more than two or three people on a podcast, it can be hard to find a time to manage.
- Don't forget that some shows need more work than others. For example, podcasts about fiction usually need more editing than podcasts about interviews.

Here are some common types of podcast:

- Interview podcasts
- Co-hosted podcasts
- Nonfiction script
- News roundup
- Podcasts that teach
- Fiction script

Podcast episode length

- Only make your episodes longer than they need to be.
- Some podcast episodes are shorter than five minutes, while others are about if a short audiobook. The podcasts can range between 10 minutes to 60 minutes

Podcast releasing schedule

- You can do podcasting as your full-time job or as a hobby. You could publish every day or once every two years, like Serial.
- Publish once a week so you can get to know your audience and get into the habit of making content regularly.
- Publishing frequencies can range between 2 to 30 days.

3. Install podcast recording equipment

- You don't have to spend a lot of money on audio equipment to get good sound. Many popular podcasts have the same basic format.
- Equipment required- podcast microphones. A good podcast mic makes your sound better and makes it easier to edit. Less noise in the background of your recording means less editing on the back end.

- Before buying your first podcast mic, consider these factors.
- Dynamic mics-Dynamic microphones are best for podcasting. They catch low-frequency and low-sensitivity. This reduces background noise for subsequent editing.
- Condenser mics- have more sensitivity and catch a wider range of frequencies. Sensitivity is useful for capturing music but not speech.
- XLR -XLR or USB are two connections for podcast mics.XLR mics are flexible. XLR mics are of various types numerous and durable. XLR mics need a mixer or interface. Beginners may find this costly.
- USB mics-USB mics are cheaper and connect into your computer. Multiple persons may record using an interface. USB mics are cheap and sound excellent. Most podcasters like this mi. For example- Fifine K730 USB Condenser Microphone for Windows PC, Laptop, MAC, PS4 with Cardioid Pattern for Voice Recording, Podcast, Video Conference, YouTube Videos. Costing INR 1900/-

Some recommendations

USB mic	
Two person set up	Samson Q 2 U 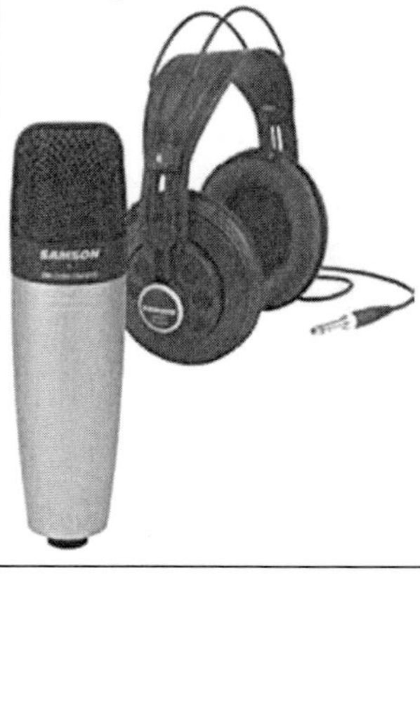

Three to four person set up	Zoom P4 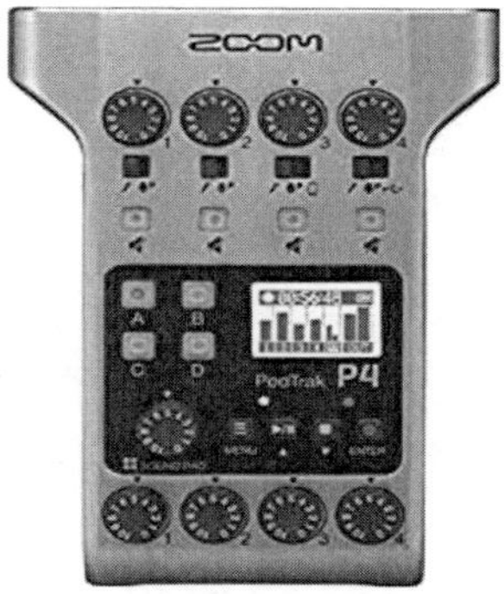
Some accessories	Rode PSA 1 Boom Arm 
	Shock Mount 
	Foam Ball/Pop Filter 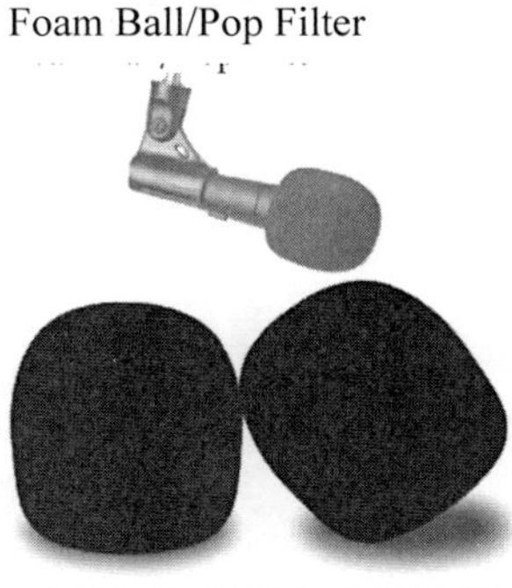

4. Select your podcast software

After recording, edit your podcast. You can organise and optimise audio here. For example, open-source software- Audacity, GarageBand.

5. Record

Your episode needs some forward planning. Three steps to follow are:

1. Structure your podcast

- Writing a screenplay or framework might help rookie podcasters avoid rambling.
- Even just jotting down bullet points can help the flow of your episode.
- Share the outline with your co-host so you can collaborate.

2. Choose a good place to record

- Record in a calm room with furniture, drapes, rugs, or carpet. Eliminate hard, flat surfaces to reduce noise. Wall foam panels may also absorb sound.

3. Install your recording equipment and software

- Connect your mic to a computer or recorder.
- Plug a USB mic into your computer's USB port. With multiple hosts, you can use an audio interface, connect each host's mic to the computer.
- Connect an XLR mic to a USB interface or external audio recorder like the Zoom P4.
- Once hooked in, pick audio input in your audio editing programme.

Microphone technique

- For the clearest sound, place your mouth 2" to 4" from the microphone.
- If you notice too many plosives or harsh sibilance in your recordings, the simplest solution is to move off-axis from the microphone.
- Set your mic up to the side, angled slightly toward your mouth. This adjustment prevents bursts of air from hitting the capsule and results in Moving off-axis from the microphone may reduce plosives and unpleasant sibilance.

Test-track recording

- Record yourself at a constant volume in various positions. Listen to the recording to find the best audio quality.
- Mistakes, stutters, and silence are okay. Next, adjust them during editing.

6. Edit audio recording

- Podcast editing focuses on audio clean-up and structural adjustments. Simplify. Overproducing will overwhelm you.

These four strategies will simplify your editing.

Create an introduction

- Your podcast introduction should inform your listeners about you and your programme. Your introduction should attract listeners and persuade them to finish the show. You may download free royalty-free music or buy your show's entrance music from these places for example: soundstripe, audio jingle, storyblocks.
- Talk as if you are talking to a single person. Use more of personal pronouns like You and Me.

Remove distractions before content

- First edit for content, then noise or other distractions.
- If you concentrate too early on noise, you may polish later-removed areas. Content first, then audio.

Fade tracks to reduce noise

- When you cut audio, you may hear pops or clicks.
- If clips start or finish abruptly, use the fade tool.

Make a list

- If you don't want to relisten to a complete programme, make a list of audio diversions.
- This list saves time and prevents mistakes.

Export audio file

- After modifying your podcast, export it to the recommended format as suggested by hosting sites for example: Anchor, Buzzsprout.

Create podcast's cover sketch.

- Potential listeners see your podcast cover image, so create an interesting cover sketch. A podcast's cover picture attracts listeners. You will require square image, JPEG files.

Tips for podcast artwork

Follow these strategies to design podcast artwork that attracts new listeners.

- Visualize podcast topic. Your artwork should tell potential listeners about your podcast. Symbolism and ambiguity turn people off.
- Design for a variety of sizes. Your artwork will be used across all media and platforms; therefore, it must look nice at 55 × 55 pixels.
- Do not overuse words or typefaces. Use no more than four words on your cover image and ensure that it is readable even at tiny sizes. Additionally, you should only choose two typefaces that complement one another.
- Avoid podcast visuals. Just like DVDs and televisions should not appear on movie covers, your podcast cover art should not depict podcasting equipment. You need not remind listeners that they are listening to a podcast!
- Use uniform branding. The artwork for your podcast is a component of your entire branding. You may make minor changes to artwork to accommodate different layouts and pages, but the imagery should remain consistent. You can use Canva, Krita open source softwares to create a cover art.

8. Set up a podcast in Google podcast, or Spotify or Anchor

- A podcast hosting services like Google podcast, Spotify or Anchor. In your podcast's description, you should outline the information that listeners may anticipate. It's also a fantastic location to employ keywords to attract the attention of readers.
- If you have a business podcast, utilise related terms and phrases so that listeners can determine the relevance of your material.
- Write show notes for -helpful links, ancillary information, contact information and short summary of the episode.
- Every podcast should have a website where your podcast lives.
- Visual Soundbites are video clips of your audio animated with your podcast artwork image and waveform. These are wonderful strategies to advertise your podcast on social media and stand out in someone's feed.
- Get your podcast listed in podacast directories for example -Submit podcast to Spotify, google podcasts, podcast index, amazon music, audible, stitcher, pandora, tunein, iHeart radio, TuneIn Radio, Listen Notes.

9. Launch and expand your show

- You have two alternatives for launching your podcast when the time comes. Choose the one that feels best to you. Promote weeks before the launch.

- Hype your show before launch.
- Share it with friends, family.
- Share podcast on social media.
- Promote others in your podcast. Podcasts make money because of ads.
- Anyone can take action to create content on podcast. You merely need enthusiasm and something of worth to provide your audience.

Activity

1. Create a google/anchor podcast on agri-preneurship.

Notes

11

To Study Script Writing in Print and Electronic Media

Learning outcome

- To learn writing for print media such as newspaper, magazine, books.
- To have in knowledge about script writing in electronic medium such as using mobile apps- LinkedIn, Twitter, YouTube and Podcast

What is Print Media?

It's a sort of mass media that publishes news and information. Print media is the oldest way to reach a large audience. It exclusively distributes information in physical copy

Books, magazines, and newspapers are popular print media. These may consist of chapters or themes, pictures, advertising, printed journals, pamphlets and so on. Print media cannot do live reporting, discussions, or shows.

What is Electronic Media?

It's a sort of mass media that distributes news and other information over the internet. Electronic media is an enhanced way to share data/news. Any user may see this sort of material once the publisher uploads (or broadcasts) it through electronic devices). It's more user-friendly than print. For example: mobile app news, desktop screening, podcasts, live discussions, live updates, live reporting, etc. The primary electronic media sources familiar to the general public are in the form of a news channel, radio presentation, video recordings, audio recordings, multimedia presentations, CD-ROM and online content.

Considerations for writing for print media

Audience

- Before writing, identify your audience.
- What degree of technical jargon to use?
- Utilise apt technical terminology for experts.
- Use simple language for novices.

Visual elements

Place picture at right places to help readers.

Provide an Overview

- Whatever you are writing, add an introduction.
- Introduction is an overview to explain – the overall purpose of the content, its usefulness to the reader.

Organising Material and Major Sections

- Create an outline.
- Research and collect relevant material and divide them into section and sub-sections.
- Divide your sections by goals.
- Use sans serif fonts in the larger size, weight and possibly colour for headings and subheadings

Provide an Index

- If you are writing a longer book or a manual include an index to help readers find information
- An index is a list of all the topics or keywords and includes the page numbers on which those words appear.
- Include Table of content

Aim for Clarity

Steps

1. Define the subject
2. Identify the chapter heading:
3. Organize the chapter sequence
4. Develop the chapter topics
5. Arrange the chapter topics in logical sequence- Use short sentences, active voice and strong verbs.
6. Organize the reference material
7. Select a working title
8. Set a writing target

Print articles must be concise yet detailed. Print media may use more complex sentence structures and word selections than radio or TV. The article must nevertheless captivate the reader. Print journalism utilises a style that appeals to the most people since it's written for wide audiences. Concentrate on the most important facts due to limited space. Print journalism generally features linear narratives. It is meant to inform and educate readers.

Writing for the Electronic media

1. Writing for Ear

Image 1: Writing for Ear

Format of radio scriptwriting

Standardizing radio format helps everyone comprehend the project. Words spoken into the microphone should be distinguished from others. Several methods work:

i) Type the speech in upper- and lowercase letters. Character names, reading instructions, descriptive text, music, and sound effects should all be typed in capital letters.

ii) Put parentheses around any non-line content, such as reading instructions, music, sound effects (SFX), notations, etc.

iii) Underline music and SFX to highlight them.

iv) It is also important for your radio script to be legibly typed and double-spaced. The standsard format for radio script writing is as follows:

MACBETH: His shadow mocks me.More shall they say, but I'm determined on knowing the worst.

MUSIC: Transition

SFX: WEIRD SOUND.....THUNDER....WIND

Witches: Abra ka dabra let us mix the ingredients

First Witch: Heart for a new....

Second Witch: And a tool of a bat

Third Witch: Wool of a rat

Witches: Fire burn and toil and trouble.

Use of software for recording

You can make use of free open-source software – Audacity .Audacity is a software application for editing, mixing, and applying effects to digital audio recordings.

2. Scriptwriting for Twitter

Twitter is a social media terrific platform to exchange ideas and grow your brand. Tips on how to write attention-grabbing tweets.

Simple Tips to Get You Started

These basic suggestions will help you learn how to write for Twitter. By being engaging, communicating directly to your readers, and using Twitter's viral nature, you may boost your brand and make your firm stand out online.

Twitter's benefits

Twitter overlaps all technologies and content. Twitter is a social network built to cater to mobile users and enjoy maximum reach.

140-Character to write

Twitter uses concise words.

Follow these guidelines to create engaging Twitter content.

1. Engage

Twitter doesn't allow much talk, so make your content compelling. Since tweets are only 140 characters, the best way is to speak directly to your audience. When your Tweets are personalised, they'll capture your readers' interest and last longer than Twitter's 18-minute limit.

2. Use shortened URL

A regular URL will use your Tweet space, but a reduced URL may attract Twitter followers to your site. You can use websites- Bitly or Buffer to tweet shortened URLs.

3. #hashtags

Twitter's framework is hashtags. Add a hashtag to boost your tweet's performance. Hashtags may brand your tweet and help other Twitter users locate and connect with you.

4. Add an image

Add a photo, video, or GIF to your tweet to boost engagement.

5. Talk to people rather than at them

Your tone on Twitter will determine whether your Tweets welcome or repel your audience. Talk with them, not at them, to maintain your tone nice. Thinking of Twitter users as friends and family might help you build a kind voice. Twitter promotes leads and engagement.

6. Add your sense of humour to Tweets

Use appropriate and professional humour enough to be shared online without compromising your business or brand.

7. Incorporate viral words

Twitter relies heavily on viral words and phrases that tell people what to do and when to do it.

Words	Used
the	1066
a	866
this	731
to	698
you	588
is	501
of	437
in	431
and	394
it	389
what	333
i	281

Source: The Most Popular Words in the Most Viral Headlines (Buffer.com)

3. Scriptwriting for a YouTube video?

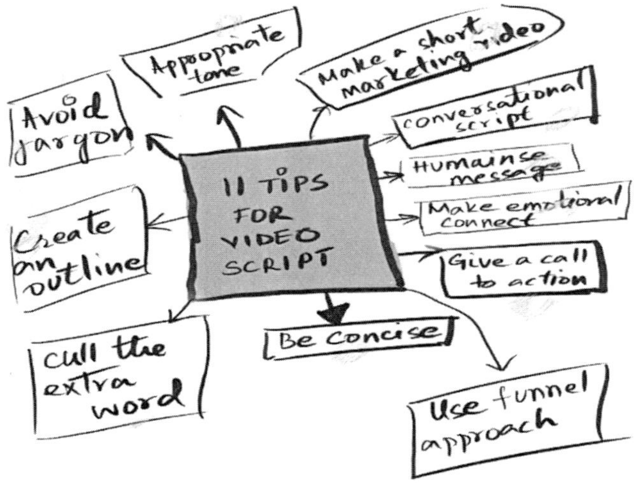

Image 2: 11 tips for video script.

Before you start video scriptwriting, answer the following questions-

- What is the goal of your video?
- Who is the target audience?
- What is your video on?
- What will viewers learn?
- What will the video lead to?

Making videos lets you team up with specialists and authorities from all over the world. You can arrange interviews, debates, or panel discussions to provide valuable knowledge to your audience.

5 Golden Rules

- Make sure your video is of 5 minutes or less and make sure it teaches something specific.
- Be friendly, talk as if you're having a conversation and sound excited.
- Ask questions and include activities that let your viewers learn by doing, like solving problems.
- Use pictures or videos to help explain your points better.
- Be creative and use interesting stories, people or images to help tell your message.

Common video and shot types

There are several common video and shot types used in filmmaking and videography. Each type of shot can convey different meanings and elicit specific emotions from the viewer. Here are some of the most commonly used video and shot types:

Video types	
Piece to Camera (PTC) 	In this style, the presenter speaks directly to the camera, giving the impression of speaking directly to the audience. This type of presentation typically involves one person speaking and is commonly used to communicate information.
Interview 	In an interview, one speaker talks to one or more other people, however it can typically be one-on- one. A common format would be the interviewer asking the questions to interviewee.
Vox Pop 	The term "vox pop" comes from the Latin phrase "vox populi," meaning "voice of the people." In these interviews, a presenter or cameraperson approaches random people on the street to get their opinions on a topic. It's important to remember that you need to ask for permission from those being interviewed if you plan to include their image in your video.
Main shot types	
Wide angle 	This technique displays the entirety of a setting and serves as a useful method to introduce a fresh location or start a new aspect of your story.
Group Shot (GS) 	This type of shot displays an entire group of people and helps the viewer to comprehend each person's position in relation to the others. It provides an overview of the entire group in a single frame.

2 Shot (2-S) 	This type of group shot captures a particular number of people and their position relative to each other within a given space. It can also be referred to as a "3 shot" or "4 shot," depending on the number of individuals involved in the frame.
Long Shot (LS) 	A long shot typically shows one or more people from head to toe. It's essential to include the entire body and avoid cutting off their head or feet in the frame, as this can appear odd to the viewer.
Mid Shot (MS) 	This shot captures a person from the waist up, providing context to the surroundings and introducing them to the viewer. It allows for a glimpse of the environment in which the person is situated.
Medium Close-Up (MCU) 	A medium close-up shot is similar to a mid-shot but captures only a person's head and shoulders. This technique is commonly used in news and documentaries when someone is speaking as it appears life-sized on TV, making it easier for the viewer to connect with them. To achieve proper framing, the bridge of the person's nose should be positioned about two-thirds of the way up the frame, leaving enough "headroom." If the person is looking outside of the frame, space should be left in front of them, and the bridge of the nose should be positioned around the center of the frame horizontally.
Close-Up (CU) 	This technique involves framing a person's entire face in the shot, allowing the viewer to observe their complete emotional reaction to a particular subject.

Big Close-Up (BCU) 	This type of shot is not frequently used and is known as a close-up shot, which provides an intense focus on a person's face or their reaction to a particular situation.
Other shot types	
Cut Away (C/A) 	This is a technique called "cutaway," where you transition from your main subject, such as a person, to another angle or piece of footage. Usually, the audio track from the previous shot continues in the background. For instance, you might cut to a close-up shot of an object the person is holding or to footage related to what the person is discussing. As an example, if you are conducting an interview about deforestation, you might cut to stock footage of trees being cut down in the rainforest.
Over the Shoulder (O/S) 	This is a filming technique called "over the shoulder" shot, where the camera captures the action from behind one person's shoulder, providing the viewer with the other person's perspective. For example, in a conversation between two individuals sitting across a table, the camera might show one person's view of the other person, including their shoulder and the back of their head, to depict their perspective of the conversation.

Understanding these common video and shot types can help filmmakers and videographers to convey their message effectively and evoke the desired emotions from their viewers.

Create a YouTube Script Outline

Outline might look like-

Introduction + hook

Hook your viewers by appealing to viewers' emotions, give anecdotes and information. The idea is to convey the message what your viewers can get by watching the video.

Introduce Yourself and Your Company

Tell viewers about yourself to build rapport. Share briefly what your organisation does. For example:

My firm, firm name, is developing a specific provision to help a certain audience fix an issue using a hidden ingredient. We strive in the expanding specified market, which was a last year's specified value market. We share characteristics.to competitor1 and competitor 2,but we have one key difference. Presently we have product's present status, team or company. We are seeking for a task that will assist us.

Create the main script

Now, focus on video subjects and write your script. While writing, consider these video script strategies:

- Visualize: Video may not always concentrate on narrator. You may zoom on goods, infographics, or switch sceneries.
- Conversational: You must converse with viewers. Try addressing the audience more, asking questions, and sharing relatable facts or tales.
- Positive tone: Your video's nature will decide its tone. Use a genuine, optimistic tone that effectively conveys your video's content.
- Limit technical shortcuts: Not everyone understands technical words. They may struggle with figures and basic facts. You should offer context for the data and explain its advantages and relevance.
- Don't overlook anything: Write out everything you want to say in your video. Say it first, what you cannot say clearly, you cannot write clearly. You may be an expert on the subject, yet you may overlook an essential point you want to make. Scripting everything minimises the amount of subsequent redos. Cull the extra words, watch for flow of script.
- End with a Powerful Call to action (CTA)

Your video script should conclude with a compelling CT. For example: likes, shares, subscriptions and downloading. Provide them infographics.

After producing the video script, verify the following –

- Will your viewers be able to understand the concepts?
- Does your script meet the demands of the audience?
- Have you given details?
- How easy is your vocabulary?
- Was your tone acceptable?
- Was the CTA apt for your audience?

The answers to the questions will assist you in adjusting your screenplay for maximum audience impact.

Narration

It is crucial that you rehearse the script by reading it aloud numerous times. By rehearsing, you will get used to the tone, pauses, and modulations necessary by the script's flow. Take feedback from your friends and family and improve.

Promote your video

A great video deserves promotion. Use Viboom to promote your channel.

4. Scripting For LinkedIn

- LinkedIn post is distinct from an article.
- It is a brief work of around 3000 characters. It might be a status update, a poll, a video, an announcement, an event promotion, or a quick message.
- A LinkedIn article is comparable to a lengthy blog post. You have 120,000 characters to both inform and amuse your audience. As it displays on search engine results pages (SERP), you may utilise it to promote your business and increase website traffic.
- Utilizing a mixture of LinkedIn articles and posts is a fantastic idea. Attempt to publish once every day, if possible. And attempt to produce one article every month.
- Numerous individuals are disheartened by the lack of views, comments, shares, and likes on their LinkedIn postings.
- Don't attempt to make a post become viral. Create content that your target client wants to read.

Some tips

1. Use plain text

Answer questions like-

- What's the goal of your post?
- How will it benefit your connections?

2. Use emoticons

Adding emoticons to your LinkedIn posts might help them stand out by showing your personality, end monotony of text. Do not overuse them.

3. Make impressive opening statement

To get attention to your article use narrative hook by use of statistics, quotes, questions, humour. Do not use hashtag, instead focus on powerful punch. Start with a story which has

1. Problem,
2. Solution, and
3. Moral

Add pictures.

4. Write single sentence paragraphs

Single phrases are easier to read than walls of text. Use three or four clear paragraph breaks between your heading and the first sentence of your text. This creates curiosity. Because the user cannot see what comes next, they must click the "see more" button. If your tale has tips, you should use bullets, numbers, or emoticon to draw attention to them.

5. List @connections or people who can help or influencers

A wonderful method to attract additional readers to your article is to @ mention someone, especially a noteworthy one. Additionally, you may @ mention a connection who has impacted your organization or career. Perhaps they provided sound advice. Perhaps they published a book on you or featured you on their show. You want them to remark on your post. Your LinkedIn post will then display on their connections feed. It's another method to develop your network and thank people who've supported you.

6. Give a CTA

Give specific instructions to encourage audience engagement. For example, to comment, like, or ask a question. You are starting a dialogue and a relationship.

7. Question

You can close posts by asking a question. By doing this you encourage comments. Comments on LinkedIn postings are rewarded. More-commented posts trend more often. This helps in building and growing network. Asking questions may stimulate discussion and fresh writing ideas.

Try these

- Is your commitment self-sustaining?
- What is your favourite tool for productivity?
- How do you self-motivate?
- What is your preferred podcast?
- How do you build relationships?
- What is one thing you wish you had known prior to beginning your career?
- Who would you most want to meet?
- What is your best networking advice?
- How do you create content?
- What is your personal motto?

8. Create intellectual property (IP)

IP is the stuff you have generated based on your professional expertise. Consider it the "best choice" for doing anything. It is founded on facts and outcomes.

You have a plenty of possibilities. If you have anything of worth to give, feel free to publish it on LinkedIn. You may be surprised by how many of your contacts are interested in gaining more insight.

9. Add apt industry related #hashtags

Hashtags are a fantastic method to connect with people outside of your network. You should come up with a hashtag that represents your brand and use it in every post. You should also choose a mix of niche and popular hashtags. Just type a word or phrase into the search bar to find them. LinkedIn also suggests hashtags for your post based on what's in it. I recommend using no more than five. Vary hashtags. Do not use the same hashtags for each post.

Activity

1. Create an informative LinkedIn post about the field activity that you undertook.
2. Create a YOUTUBE video regarding agricultural topic.
3. Create a podcast on agriculture.

Notes

____

____

____

____

____

____

____

____

____

____

____

____

____

____

____

12

To Study the Script Development for Radio and Television

Learning outcomes

- You should know how to make an audio or video script.
- Describe the differences between the different kinds of programmes.
- Examine the different script development stages.

Before we start, please complete the following activity.

Reflect on the programmes you watch on TV or any other programme using Over the top (OTT) platforms like amazon, prime video, Netflix etc. How many of it you classify as

1. Entertainment
2. Educational
3. Informational

A programme or film should satisfy two broad requirements to be termed educational:

i) It should be methodical and structured.

ii) It should seek to increase the receivers' knowledge permanently.

Reflect on this for a bit before moving further. The material in newscasts is not presented in an ordered fashion but instead jumps from topic to topic based on current events.

Quiz shows deliver knowledge in an unfocused, random manner. In a way, the information offered by newscasts is strictly temporary since it is only valid until the next news programme. Therefore, the information provided by quiz shows is of little or no significance. Thus, the broadcast and quiz programmes do not satisfy the abovementioned standards.

We would categorise news broadcasts as informative, and quiz shows as entertaining. A more accurate term for quiz shows would be "infotainment" - a combination of knowledge and entertainment.

A medium carries a variety of popular art forms or genres. In electronic media, these categories are often referred to as shows or programmes.

A format describes the design of a script., The exceptional standards (of writing) are used while scripting with specific reasons.

While writing a script, you must consider style, medium, and genre. An excellent script is distinguished by the following:

- Variety of shots
- Different perspectives
- Interviews weaving in and out to make a dialogue
- A good balance between what is said and what is shown.
- The script is essential to any successful show, whether on stage, radio, film, or videotape. It is a carefully thought-out plan that tells actors, directors, and technicians what to do.
- Writing styles- Keep the writing style simple, fluent, lively, and natural.

Different types of formats

Writers and producers of media programming employ a range of presentation forms for their broadcasts. Radio, television, cinema theatre, and print media share most of these forms. Radio and television have modified these forms for improved communication and impact. Producers and authors of educational programming often use a single format or a mix of two or more formats that best meets the needs of a specific topic or subject. A writer must focus on selecting an appropriate format for successfully presenting their program's content from the earliest stage of programme creation. Decisions on the layout are often influenced by the nature of the subject matter, its needs and goals, the audience's interests, and, most crucially, the availability of adequate production resources and the audience's interests.

Formats for programming

The following are radio and television formats:

i. **Story/straight talk:** The straight talk or story approach, better suited to the radio but often used on television, typically employs a single narrator. This genre includes illustrated speeches, special reports, newscasts, and presenter-graphic-presenter programmes. To give a slight variation, two people/voices are occasionally employed with suitable images to deliver information. In addition, radio and television often use fully scripted tales and talks to offer a range of instructional information.

ii. **Interview:** In this approach, the interviewer and interviewee participate in face-to-face conversation. The interviewer engages the interviewee in conversation on behalf of the viewers or listeners and strives to extract from the intriguing subject information pertinent to the issue. This interview style is ideal for presenting current events or informational and instructional programmes in which a live scholar, scientist, or expert addresses the audience, improving the program's interest and legitimacy. Interviews are not predetermined. Only the opening and closing questions and speeches are prepared in advance and sometimes rehearsed. The rest of the show is improvised and unscripted, with much depending on the interviewer's self-assurance, competence and preparedness. Relevant questions structured with clarity, organised rationally, and posed with confidence, tact, and spontaneity are the foundation of a successful interview.

iii. **Panel discussion:** This structure is an extension of the interview. Here, a panel of at least three experts examines a controversial topic of public interest. A panel brings together individuals with divergent beliefs and perspectives in a one-on-one forum. Discussions are directed and moderated by the panel chairperson. There is no need for a script, but the moderator or host must prepare extensively. To guarantee a friendly, orderly, and substantive debate, s/he must be vigilant, well-informed, discreet, and in complete control of the situation.

iv. **Feature:** In this style, a single topic or issue is selected to present its numerous facets and debate of facts, views, beliefs, arguments, disputes, and counterarguments - all about the same topic. For instance, an in-depth documentary about cave art may be created, covering several facets of the topic.

v. **Magazine:** The name magazine derives from the print media. Similarly to how a weekly or fortnightly magazine section offers articles on various themes, a radio or television magazine show might have various unconnected issues. The distinction between a magazine and a feature is that the former covers multiple topics, sometimes aimed at different audiences within the same show, whilst the latter focuses on diverse facets of the same subject. A magazine programme is comparable to a visit to a fair or department shop when many items are available for visitors or audience groups. Due to the diversity of its content, the magazine show can maintain the viewers' interest for longer.

vi. **Drama:** Primarily a literary genre of narrative, Television and radio have directly borrowed this format from the theatre. Drama is a highly engaging and enjoyable genre that is particularly successful when meant to appeal to the audience's emotions. Drama allows us to display and observe individuals, their behaviour, temperaments, and traits in realistic or lifelike settings. Writing and creating a quality television drama is challenging. Numerous social science and historical themes are well suited for theatrical adaptation and presentation. Drama demands realistic settings and a cast of performers to represent circumstances and happenings.

vii. **Documentary:** In contrast to drama, the documentary genre takes an objective approach to its subject. Here, an issue is examined in detail, and its facets are factually examined. A documentary focuses primarily on capturing actual life, including genuine people, real events, real noises, and fundamental human experiences. A documentary filmmaker's primary responsibility is to study and investigate the topic's numerous elements extensively. S/he uses a variety of techniques to present their materials, including narration, montage, interviews, discussions, eyewitness accounts, candid visuals, and shots of actual events, as well as comments from the "man on the street" and an expert's opinion, which are all arranged logically. The documentary style is appropriate for presenting current events, particularly social and political ones.

viii. **Quiz:** This style is ideal for delivering educational and current affairs information. The quiz style is very engaging and interactive. It maintains the viewers' interest and keeps them immersed and captivated by the presentation.

Depending on the discretion of the show writer and producer, these formats may be utilised in combination or independently. Additionally, it relies on the type, aims, topic matter, available time, money, and resources.

Radio script format

- The page should be divided into two columns, with instructions on the left and dialogue on the right.
- The names of the characters and instructions for music, sound effects, and other things are all written in all CAPS in the left-hand column. The only thing not written in capital letters is dialogue and single-spaced.
- Only the right column contains dialogue. Dialogue is always written with double spacing between lines and upper and lower case (like typical prose). This is done to make it different from the instructions. In addition, this lets the cost and production crew know quickly what is and is not dialogue.
- The directions for how to read a line are in parenthesis and ALL CAPS on the right side of the script, in the appropriate places.
- Not only are the directions for music and sound effects written in ALL CAPS, but they are also underlined.

Table 1: Format of radio script

Instructions	**Dialogue**
Name of Character	Dialogue is always double spaced and written in upper and lower case
Direction	
Music	
Sound Effects	

Scriptwriting for Television

It is primarily a group activity. It could be a conversation between the scriptwriter, the producer, and the others on the production team. This conversation is primarily a joint effort that starts with the programme ideas that are finally made. The scripting process is all about creativity, imagination, and working together.

TV scripting is a process that involves several steps or processes that don't have to happen in order or be separate from each other. These steps run together. Here's how to do it:

- Programme idea
- Programme outline/brief
- Programme research
- Message planning/programme design
- Programme structure and visual treatment
- Storyboard making
- Draft-script
- Final script
- Camera script

Here is a detailed explanation of each of the above steps in writing a script. This will help you understand it better.

Steps in the process of writing a TV script

Programme idea: Talk about the concept and judge it based on the criteria for choosing media for video production. If it meets the requirements, finalise and accept the idea so it can be made.

Programme brief/outline: Expand the chosen and tested programme idea into a programme brief, spelling out the exact title, target audience, length, programme goals (in terms of behaviour), and an outline or synopsis of the program's content. The programme brief is the first written statement about making a script for TV or video.

Research: Plan and do thorough research on the topic and the people you want to reach. Look for pictures and other interesting, meaningful information about the topic. Check with books, encyclopaedias, journals, experts, and people who know much about the subject. Know the people you want to reach. Find out the truth about their past, interests, what they like and don't like, etc. Also, consider the viewing environment and context.

Selection of materials

1. Choose the most relevant information, resources, and ideas that correspond to your program's goals.
2. Determine and estimate the content that should be included in the final programme.
3. Remove anything that is not directly relevant to your goals.
4. Keep your goals and the duration of your programme in mind when you estimate the amount of information to include in your final schedule.

Programme design strategy/message planning: The most significant and innovative phase of programme creation is message planning. Consider and determine programme design and presenting approach at this point. Explore and develop novel methods for delivering your programme to your audience. Always bear in mind your audience's interests, the nature of your subject, your program's aims, and the visual potential of your medium. Think carefully about presentation format or a combination of formats, programme style, pace, and approach, information size and emphasis, media options, visual analogies and examples, Utilising attention-getting strategies, humour, repetition of essential concepts, visuals, animation, and other production techniques, resources/ gimmicks.

Programme structure and visual treatment: The stage at which the program's structure or form emerges. Consideration is given to the creative process of providing the programme with a distinct shape and form. Consider problems such as which sequences will comprise the program's core (middle) part. Consider a decent beginning for your presentation since the opening sequence captures the audience's attention, and it may make or break the performance. Also, consider a satisfying and realistic conclusion.

Develop a comprehensive visual strategy for your show's middle sections or sequences. Consider visual representations and visual comparisons for abstract notions and essential ideas. Give your programme parts a natural and logical order, so they go from one sequence to the next.

Consider potential visual and audio linkages or transitions from one sequence to the next. Keep track of time and goals. Create a schedule indicating how much time should be dedicated to each sequence, considering the covering of the contents and the relative importance or weighting of the programme goals and essential teaching points.

The treatment should describe in detail what the audience will see, hear, think, and feel.

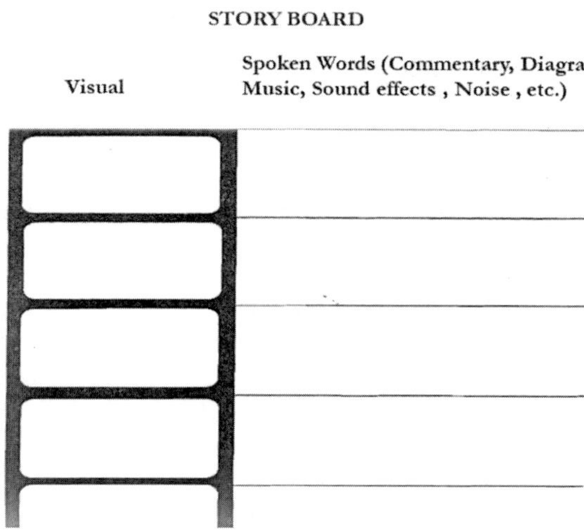

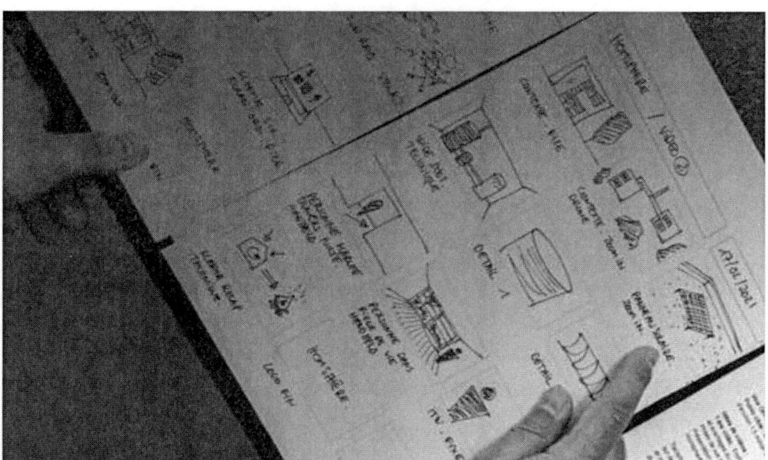

Image 1: Storyboard

Storyboard making: A storyboard is a comprehensive, shot-by-shot description of a programme written in two vertical columns on paper. The rectangular boxes (3:4) on the left column are used for drawing images/ sketches with the shot size stated on each, while the right column is used for writing accompanying text, sound effects, and music.

Before filming and editing your video, there are two crucial pre-production tasks to complete: writing a script (or preparing interview questions/answers) and creating a storyboard. These tasks allow you to effectively manage your time, equipment, and costs. Writing and practicing a script will help you present confidently on camera, while creating an engaging storyboard will ensure that your video meets the five golden guidelines we previously discussed.

You don't need to be a skilled artist to create a good storyboard! It can simply be text-based with descriptions, providing you with a clear plan for production and post-production.

It is essential to have a storyboard because:

- It compels the writer to visualise their work. At this point, the TV/video screenplay starts to take on the appearance of sequential images, with sound and action detailed alongside.
- With the aid of the storyboard, it is straightforward to analyse the general movement and development of the script.
- Using the storyboard, it is possible to analyse the general development and growth of the script. Examples include adding, removing, altering, and restructuring pictures and sequences.
- All members of the production crew would benefit from studying the storyboard.

Script development has three phases

i) Draft script: First full-length script containing all visuals/captions, conversation, music, and sound effects. Discuss your first draft with colleagues, experts, and production team members. Be willing to receive criticism and improvement suggestions.

ii) Final script: All draft script recommendations are incorporated. The final script is sent to the producer, who analyses it and creates a camera script. After this stage, the playwright usually leaves, and the producer takes control, although their contract continues until the show is completed.

iii) Camera script: This is the script's crucial stage. The scriptwriter's final document becomes the camera script for the producer and crew. It's a technical screenplay with shot division, production team instructions, and logically sequenced music effects and dialogue. A camera script outlines all programme aspects and resources. It's an essential document for the producer.

Testing of the script: Last is script testing. When producing many shows, testing the script with the target audience is essential. Script testing evaluates the program's presentation, comprehensibility, linguistic difficulty, relevancy and diversity of topics. The contents of the scripts are read aloud to the target audience to assess their strengths and flaws. Some scholars try to dramatise the plot as it appears in the script. The screenplay writer may modify the substance and arrangement of the script, as well as the design sequences and several images, based on the results of such testing. Script testing is a formative assessment process that gives rapid input on many product parts, which may be updated appropriately before production.

TV script format

- Split the page in half and put the video on the left and the audio (and dialogue) on the right. The words "video" and "audio" are usually written at the top of each page.
- ALL OF THE WRITING ON THE VIDEO SIDE IS IN ALL CAPS. In addition, there are instructions and pictures taken with a camera.
- Dialogue is always written in uppercase and lowercase letters with two spaces between them on the right side of the page, which is the audio side. The rest is written in ALL CAPS.

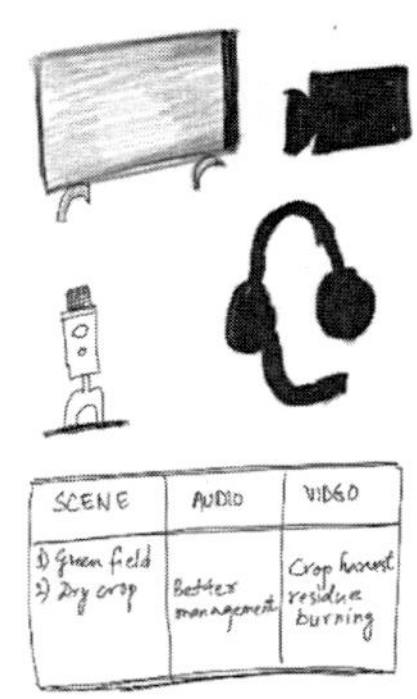

- ON THE AUDIO SIDE, CHARACTER NAMES ARE TYPED WITH ALL CAPITAL LETTERS.
- On the audio side, sound effects and music are shown on separate lines in CAPS.
- ON THE VIDEO SIDE, PHYSICAL MOVEMENTS ARE SHOWN IN CAPS.

Table 2: TV script format

Video	Audio
INSTRUCTIONS	CHARACTER NAME
CAMERA SHOTS	SOUND EFFECTS
PHYSICAL MOVEMENTS	MUSIC

How to write a documentary script

In general, there are six main parts to a documentary.

i) Narration: The narrator helps set the scene and keep the story moving from one scene or interview to the next. The amount of telling should be kept to a minimum. In a documentary, the characters should help tie everything together. They should not take over and become more important than the main subject.

ii) Voice-over: Here, the narrator is not visually shown. Instead, they deliver comments or explain what we see on screen. Voice-overs should primarily be utilised to further the plot.

iii) Interviews provide a snippet of events in a documentary.

iv) Graphics: This may include captions, diagrams, charts, photographs, and maps.

v) Music is utilised to create a positive environment

vi) Natural sound and sound effects are used to improve mood.

You can find many great free collections of stock footage, images, music, and sound online. Here are some useful websites

Website	**What it offers**	**Tips for searching**
Mazwai	Stock videos.	Search by keyword or browse by category.
Media Hopper Create	University of Edinburgh media collection.	Select 'Creative Commons' at the top of the page to see a broad range of video and audio content available for download and re-use. Filter by channel to browse by topic or course title.
Openverse	Images and audio clips.	Use the dropdown menu to view all or filter by image or audio.
Pexels	Stock images and videos.	Search by keyword or browse by category.
Pixabay	Videos, images, illustrations, vector graphics, music and sound effects.	Use the dropdown menu when searching to filter by content type.
Unsplash	Stock photography.	Search by keyword or browse by category.
Videezy	Stock videos.	Search by keyword or browse by category. Filter by file type and format, licence type and resolution.
Videvo	Stock video, music, sound effects, video templates, motion graphics.	Search by keyword or browse their collections of different themes.

If you use Creative Commons third-party content, like stock footage, you must give credit in the video credits and follow the licence terms.

- A quick way to do this is to make a black end board titled "Attribution" and save it as a template to use for multiple videos. You can use programs like Photoshop or Canva.

- Include the following information for each third-party content item in your video:

[Title], [Creator], [Licence], [Location], [link to source]

Tip: When you find and download stock clips and other content for your video, remember to save the source webpages as bookmarks. This way, you can easily access the attribution information later without having to search for everything again!

Examples

- Hampton Wind Park, Bluedawe, CC BY-SA 3.0, Wikimedia Commons, https://commons.wikimedia.org/wiki/File:Hampton_Wind_Park.ogv
- Aerial View of an Industrial Plant, Kelly Lacy, free to use video, Pexels.com, https://www.pexels.com/video/aerial-view-of-an-industrial-plant-2406491/
- This example shows stock clips attributed in a video created for one of our short online courses.

Free editing software

- Now that you have filmed your footage, you can begin to chop, cut, and combine it using a variety of different editing software. We mostly cover free options here, but also mention some paid-for software you might already have access to, or may want to consider investing in.
- iMovie
- Platform: macOS, iOS, iPadOS
- Shotcut- Platform: Microsoft Windows, macOS, Linux. It is a free, open-source video editing application that runs across most operating systems, including Linux.
- Find out more information on Shotcut, as well as links to tutorials on its use, on the Shotcut website.
- Davinci Resolve- Platform: Microsoft Windows, macOS, Linux
- Davinci Resolve is a powerful free video editing software that's popular among film industry professionals for its colour grading capabilities. It has many features, but may require a higher-specced computer to run smoothly. The software is free because it's made by BlackMagic, who also offer hardware products. Advanced features like HDR grading require the paid-for version, but the free version is enough for most editing needs. You can find a list of features and tutorials on the software's website. BlackMagic website.

Case study- Distance education as an example of script development stages

In distance education, for instance, developing an audio or video program script may be seen in three separate phases.

- Primary research and brief programme development
- Elaboration of the program brief into academic note
- Transformation of academic notes into a script

Program Brief

The programme outline is the starting point for the actual production of an audio or video show. The programme summary provides information on-

- The target audience
- Working title
- Objectives
- Content outline
- Any additional information

The academic note

The academic note is a literal and expanded version of the overview of the content. This includes suggestions for visual content, sound effects, places, and pricing. You may characterise Academic content using the following phrases:

1. Media content
2. Learners' characteristics viz., age knowledge, commitment cultural conditions
3. Teaching objectives
    i. Effective
    ii. Motivational
    iii. Experiential
    iv. Cognitive
4. Format and style
    i. Studio or location or mixed
    ii. Presenter in vision or out of vision
    iii. Documentary docu-drama
    iv. Demonstrated or illustrated or lecture or case study
5. Outline the content

Activity

- Create a video script for hydroponics.

Notes

References

Anja Christnick and Martina Padmanabhan (eds). Cultivate diversity. (2013). Margraf Publishers GmbH.

Berkenfeld, D and Silverman, L.(n.d.) Easy Macro Photography with the AF-S DX Micro-NIKKOR 40mm f ... - Nikon USA. https://www.nikonusa.com/en/learn-and-explore/a/tips-and-techniques/easy-macro-photography-with-the-af-s-dx-micro-nikkor-40mm-f-2-8g-lens.html

Bonk. Lawrence.(2022).Audacity. Free Audio Editing App Audacity Gets a Makeover With a Huge Feature https://www.lifewire.com/free-audio-editing-app-audacity-gets-a-makeover-with-a-huge-feature-rich-update-6743936

Buzz sprout.(2021). How to Name Your Podcast https://www.buzzsprout.com/blog/names-for-podcasts

Buzz sprout.(2022). How to Start a Podcast: Complete Step-by-Step Guide. https://www.buzzsprout.com/blog/how-to-start-a-podcast

Canon EPF lens. (n.d.) EF-S18-200mm f/3.5-5.6 IS. https://gdlp01.c-wss.com/gds/2/0300003652/02/efs18-200f35-56-is-im2-eng.pdf

Club,Sam. (2022).The 7 Best Camera for Video in 2022 – Sam's Club Online Shopping. https://www.samsclubonlineshopping.com/the-7-best-camera-for-video/

Collins Dictionary.(n.d.). Workshop definition and meaning. https://www.collinsdictionary.com/dictionary/english/workshop

Computer organization - Computer Science. (n.d.) Wiki. https://computersciencewiki.org/index.php/Computer_organization

Digital Audio Fundamentals - Audacity Manual. https://manual.audacityteam.org/man/digital_audio.html

Digital single-lens reflex camera (n.d.) -Wikipedia. https://en.wikipedia.org/wiki/Digital_single-lens_reflex_camera

District Rural Development Agency - Wikipedia. https://en.wikipedia.org/wiki/District_Rural_Development_Agency

Field Level Handbook SEGA FAO. (2001). http://www.fao.org/sd/seaga/downloads/Eng/Field_Engl.2002.pdf

Global Hunger Index. (2022). India - Global Hunger Index (GHI) - peer-reviewed annual publication designed to comprehensively measure and track hunger at the global, regional, and country levels

GoldMore.(n.d). LED Flashlight Manufacturer & Supplier in China . https://goldmore.com/led-flashlight/

Hilary Sims Feldstein and Janice Jiggins (eds). (1994). Tools for the field – Methodologies handbook for gender analysis in Agriculture. Kumarian Press Library of management for development.

How to Attach a Camera to a Tripod: 10 Steps (with Pictures) - wikiHow. (n.d.) https://www.wikihow.com/Attach-a-Camera-to-a-Tripod

How to Create Video for Online Courses - Grades | Coursera. How to Create Video for Online Courses - Grades | Coursera

Instructional Technology Resource Centre. (n.d.) https://www.uwo.ca/its/itrc/about/hardware_and_software.html Western University.

Jomon Mathew and Joby Varghese (2011). Non-Governmental Organisations in India , Yojana , November.

Levine, Nicole.(2022). How to Connect Speakers to Your Laptop – wikiHow(n.d.) https://www.wikihow.com/Connect-Speakers-to-Your-Laptop

Lucid.(n.d). Glossary of Meeting Terms - Lucid Meetings. https://www.lucidmeetings.com/glossary

Mahesh,V., Swathi, P.S., Lekshmi, Dilip Ananda Pawar , Daliyamol, Kumar, Alok and P. Prakash.(2017). Matrix Ranking- An important PRA tool to assess farmers http://eprints.cmfri.org.in/12908/1/Swathilekshmi%20P%20S_2017_Agricultural%20Science%20Digest%20_Matrix%20ranking.pdf

Manage. (2014). Community Radio and its Operationalization. https://www.youtube.com/watch?v=6KDnvIxifYg&list=PLdyTOmBb-eOm0heGwO3y_uUae89VasrQL&index=9

Manage. (2014).Community Radio and Its Operationalization Page Number 143-148 https://www.manage.gov.in/pgdaem/studymaterial/aem102.pdf

Neela Mukherjee. (2009). Speaking to Power 27 voice tools. Building bridges for participatory learning action and policy making. Foreworded by Robert Chambers. Concept Publishing Company.

Paar, Morgan. (2017). Fresnel Lights: How They Work and Why They Should Be in Videomaker. https://www.videomaker.com/article/c03/19113-fresnel-lights-how-they-work-and-why-they-should-be-in-your-arsenal/

Plaejaro, G.L. (2014). How to Conduct a Panel Discussion - slideshare.net. https://www.slideshare.net/larryprats/how-to-conduct-a-panel-discussion-37521531

Projector – Wikipedia (n.d.) https://en.wikipedia.org/wiki/Projector

R. Sooryamoorthy and K.D Gangrade (2006). Non-Governmental Organisations in India : A cross sectional study , Rawat Publishers, Jaipur.

Ryan, Cory and Hall, Kyle. (2022). How to Clean a 35mm Film Camera and Lens - wikiHow. https://www.wikihow.com/Clean-a-35mm-Film-Camera-and-Lens

Screenwriting Transitions: The Ultimate Guide. https://freshmenscreenplay.com/transitions-in-scriptwriting-the-ultimate-guide/

Shubham Soni. (2017). Participatory rural appraisal(PRA). https://www.slideshare.net/ShubhamSoni127/participatory-rural-appraisalpra

Small group techniques. (n.d.). http://ecoursesonline.iasri.res.in/Courses/Extension%20Methodologies%20for%20Transfer%20of%20ag.%20Tech/Data%20Files/lect10.html

Socratic method- https://www.youtube.com/watch?v=X1VcFggcLTY

Somesh Kumar. (2012). Methods of Community participation .A complete guide for practitioners. Vistar Publications.

Steve Laube.(2014). How to Write a YouTube Script: Video Script Example and Tips - Viboom. https://viboom.com/en/academy/how-to-write-a-youtube-script-video-script-example-and-tips/

Steve Laube.(2014). What is One Thing You Wish You Had Known? - The Steve Laube Agency. https://stevelaube.com/one-thing-wish-known/

Syeda Hameed .(2011). A commitment to the Voluntary Organisation, Yojana, November.

Tamoor Zulfiquar (2015).Non Projected AV Aids https://www.slideshare.net/TamoorZulfiquar1/non-projected-av-aids

UNDP. (2022). India ranks 132 on the Human Development Index as global development stalls | United Nations Development Programme.

USAID (2011). Developing and using case studies MEAS TN Developing and Using Case Studies - Gunderson - UF - Aug 2011.pub (illinois.edu)

USAID (2011). Presenting workshops to adults available at:MEAS TN Presenting Workshops to Adults - Myers - UF - Aug 2011.pub (illinois.edu)

What is an index and do you need one? - Editor Group. https://www.editorgroup.com/blog/what-is-an-index/